Energy Efficient Transformers

Energy
Efficient
Transformers

Barry W. Kennedy

McGraw-Hill

New York   San Francisco   Washington, D.C.   Auckland   Bogotá
Caracas   Lisbon   London   Madrid   Mexico City   Milan
Montreal   New Delhi   San Juan   Singapore
Sydney   Tokyo   Toronto

Mcgraw-Hill

A Division of The **McGraw·Hill** Companies

1 2 3 4 5 6 7 8 9 0   DOC/DOC   9 0 2 1 0 9 8 7

ISBN 0-07-034439-6
Part of 0-07-034288-1

The sponsoring editor for this book was Harold B. Crawford, the editing supervisor was Scott Amerman, and the production supervisor was Tina Cameron. It was set in Palatino by Estelita F. Green of McGraw-Hill's Professional Book Group composition unit.

Appendix L, "DTCEM User's Manual," originally appeared in its entirety as a manual published by the U.S. Environmental Protection Agency (EPA). Copyright © 1997 by the U.S. Environmental Protection Agency. Used by permission of the authors and the publisher.

Printed and bound by R. R. Donnelley & Sons, Inc.

McGraw-Hill books are available at special quantity discounts to use as premiums and sales promotions, or for use in corporate training programs. For more information, please write to the Director of Special Sales, McGraw-Hill, 11 West 19th Street, New York, NY 10011. Or contact your local bookstore.

 This book is printed on recycled, acid-free paper containing a minimum of 50% recycled, de-inked fiber.

The opinions contained in this book are those of the author of the book and do not necessarily reflect the opinions of the author's employer, the Bonneville Power Administration, or the United States Government.

Contents

Preface

Since the 1970s, the increased cost of electrical energy has resulted in utilities requiring transformer manufacturers to build more efficient transformers, although some small utilities as well as commercial and industrial users of transformers continue to use cheaper but inefficient transformers. Today's power transformer, be it distribution or substation, is more efficient than ever. Computer design and manufacturing techniques provide the purchaser of transformers with many choices. New manufacturing techniques and new types of core materials today provide cost-effective and energy-efficient transformers to the transformer user.

The decision as to whether to purchase a low-cost, inefficient transformer or a more expensive, energy-efficient transformer is primarily an economic one. In addition, energy-efficient transformers reduce energy consumption and consequently reduce the generation of electrical energy and greenhouse gas emissions. Energy-efficient transformers cost more but use less energy than low-efficiency transformers. Before you decide what transformer to purchase, you must weigh the capital cost of high- and low-efficiency transformers against the cost of their losses over time. You also must size the transformer so that it will operate efficiently. I have designed this book to provide you and other transformer purchasers with the knowledge and tools that will help you decide what transformer size and efficiency ratings are best for you.

There are many books about transformers, but none is dedicated solely to the efficient use of transformers. My goal in writing this book is to help users of transformers save money as well as energy. If you are new to transformers, you will find that Chap. 2, "Transformer Characteristics," will provide you with the necessary fundamentals on transformer theory, construction, operation, and energy consumption. If you are familiar with transformers, you will find the later chapters on new transformer technology, such as amorphous and laser-etched metals, as well as insights into specially designed transformers, such as K-factor transformers, informative and valuable. And for both the beginner and the advanced user of transformers, I have provided a comparison of various methods for evaluating the cost-effectiveness of high-efficiency transformers, taking into account the savings in energy over time. I also discuss in great depth transformer losses and how to reduce them in Chap. 3, "Transformer Efficiency."

I have organized this book into logical steps to help you obtain easily the information you need to purchase and operate transformers cost-effectively. I explain how to use this book in Chap. 1.

In order to evaluate the cost-effectiveness of energy-efficient transformers, you need to know how to determine transformer losses and efficiency. Transformer losses are divided into core (no-load) and coil (load) losses. Core losses occur continuously, independent of the load, while coil losses are dependent on the load. In either case, these losses represent energy lost to heat during the voltage/current transformation process. Heat reduces both the life and carrying capability of all transformers. "Transformer Efficiency," Chap. 3, provides you with the causes of losses, how to reduce them, and how to determine them.

After you determine the amount of losses, you need to calculate the value of losses and weigh the savings in losses against the capital cost of energy-efficient transformers. The value of losses has increased significantly over the last two decades and varies from one transformer user to another. "Value of Losses," Chap. 4 will provide you with various methods to value your transformer losses. In addition to evaluating losses, you must evaluate the cost of the transformer itself.

To help you to determine the cost of transformers, I cover the various factors that affect the cost of transformers in Chap. 5, "Transformer Cost."

After you determine the amount of losses and the value of those losses, you need to evaluate the losses over time so as to decide what transformer to purchase. In selecting transformers, you need to consider the long-range operating costs as well as the initial cost of the transformer. In this book I provide a method to help you select transformers that are cost-effective over the 30-year life of the transformer. I discuss this in Chap. 6, "Transformer Economics."

During maintenance of transformers, you may be faced with a decision about whether to replace a transformer or refurbish an existing transformer. With the transformer economic tools learned in Chap. 6, I explain in Chap. 7 how to determine whether you should replace or refurbish an existing transformer.

New technologies for improving the efficiency of transformers are being developed as consumers demand more energy-efficient transformers. New low-loss transformer technologies became commercially available in the 1980s. I discuss low-loss transformers in Chap. 8. I devote Chap. 9 to amorphous metal distribution transformers. Amorphous metal transformers contain a specially manufactured metal that has 75 percent lower no-load losses than a standard silicon steel core transformer.

Harmonic losses often result in overheating of transformers. Transformer manufacturers design K-factor transformers specifically to handle harmonics. K-factor transformers consequently have lower losses than a conventional transformer. I show you in Chap. 10 how to select K-factor transformers that meet your special harmonic requirements.

The transformer manufacturing industry provides you today with many choices of transformer efficiencies. My goal in this book is to help you sort through these choices. I give you a glimpse into the future of how transformers are being made even more efficiently using new cooling methods such as cryogenics. I present these trends in Chap. 11.

Finally I include in the back cover of the book a computer program for evaluating losses on distribution transformers. This program is called the Distribution Transformer Cost Evaluation Model (DTCEM). It was developed by the Environmental Protection Agency to encourage the use of energy transformers for the purpose of reducing greenhouse gases. I also include in Appendix L a manual on how to use this program.

My purpose in writing this book is to provide you with straightforward steps for selecting and operating transformers, taking into consideration the energy use of transformers and the cost over time. With this knowledge and tools, my hope is that you will make wise decisions on what transformer to purchase. I intend to bring this information to all users of transformers, be they technical or nontechnical personnel in a electric utility or industrial or commercial organization.

My special thanks to Wayne Beaty, former Editor of Electric Light and Power, for his guidance and encouragement, Don Duckett of General Electric Company for his expertise, Larry Lowdermilk of General Electric Company for his help, Denise Knight of Allied Signal Inc. for her assistance, Paul Barnes of Oak Ridge National Laboratories for his insights, Peter Scott of the Environmental Protection Agency for his assistance, and Hall Crawford, Senior Editor of McGraw-Hill, for his patience.

You, the users of electrical transformers, benefit from the selection and operation of cost-effective, energy-efficient transformers. You have the data on the value of losses, the transformer costs, and operating time needed to make the decision that benefits you and your customers. With this book you will have the knowledge and methods for evaluating the cost-effectiveness of energy-efficient transformers that meet your need to save energy and money.

Barry W. Kennedy

Foreword

I highly recommend this book on energy efficient transformers by Barry Kennedy. His dedication to this subject and his many years of experience and expertise with the Westinghouse Electric Corporation as a transformer design engineer and with the Bonneville Power Administration as a transmission and distribution system efficiency expert more than qualify him to write a book on this much-needed subject.

The voltage in power systems is normally transformed up to three or four times from the generation source to the consumers, and a lot of energy is lost during those transformations. In this era of competitiveness and cost cutting, it's important not only to the transformer designer but also to the engineers who specify and apply transformers to use the best available technology, and this book will do just that.

This book will be of valuable assistance to all engineers, technicians, and energy managers in the proper application of energy efficient transformers. It is the most complete book that I have seen on transformers.

This book could also be an excellent book for educators and students who are studying electrical engineering and mechanical engineering. It not only covers the basics of transformer design and application but also goes into intricate detail about every phase of transformer characteristics, efficiency, costs, operation, and maintenance.

Energy Efficient Transformers will have a very long shelf life and will be a valuable addition to anyone's library who is in any way interested in or needs a good reference source on transformers. It should become the "bible" on the efficient and cost-effective use of transformers.

Wayne Beaty[*]

[*]Wayne Beaty has recently retired as Editor of *Electric Light & Power* magazine, *Power Delivery* magazine, and *Utility Automation* magazine. He is the former Senior Editor with *Electrical World* magazine and Technical Editor of *Rural Electrification*. He is still the editor of the highly rated *Handbook for Electrical Engineers* published by McGraw-Hill. Mr. Beaty also had a long and distinguished career as Chief Distribution Editor for West Texas Utilities Company, an operating company of the Central and South West Corporation. Mr Beaty also worked for the Electric Power Research Institute for many years as Manager of Member Services and Assistant Director of the Washington, D.C., office.

1

Introduction

Utilities improve the efficiency of their transmission and distribution systems by reducing losses. Industrial and commercial users of electricity also improve the efficiency of their electrical distribution systems by reducing losses. They reduce transmission and distribution losses by using the following seven basic methods:

1. Substitute larger conductors for those currently in use
2. Increase the system voltage
3. Improve system power factor by adding shunt capacitors
4. Add lines or feeders
5. Add or balance phases
6. Use energy-efficient transformers
7. Reconfigure the electrical system

While some of these modifications can be done more easily than others, the use of energy-efficient transformers is always easily accomplished. It does not take a great deal of technical expertise to correctly use high-efficiency transformers. If you use modern high-efficiency transformers, you do not need to know that they are constructed with low-loss materials using new manufacturing techniques. It is a low-tech application of a high technology. All you need to know is how to choose the most cost-effective energy-efficient transformer. Why is the efficient application and operation of transformers important to electrical utilities and industrial and commercial users of electricity?

Transformer Types and Their Energy Usage

Transformers cause an electrical power system to operate more effi-ciently. They step the voltage up so that more power can be transmit-ted efficiently at longer distances. They operate in electrical utility sys-tems at both the transmission and distribution levels. They step the voltage down in factories and commercial complexes and residences to a level at which the equipment and appliances can operate.

Transformers act as passive devices for transforming voltage and current. They operate more efficiently than most electrical devices. Transformers operate from 95 to 99 percent efficiency. However, because there are so many transformers in use, small losses in each one add up to significant amounts. Coupled with this fact, the increased cost of losses forces users of transformers to require transformers to operate more efficiently.

There are various types of transformers. The two main categories are substation and distribution transformers. *Substation transformers* (100 to 500,000 kVA) are large two-winding transformers built to step the voltage up from the low voltage at the generator to the high voltage of the transmission system. Other substation transformers are located at substations in the transmission and distribution system for stepping voltage down to be distributed to various loads, whereas distribution (.5 to 3000 kVA) transformers are smaller but more numerous than substation transformers. They are used to step the voltage down to a level that allows the electricity to be used by homes, offices, and facto-ries.

In a distribution system, distribution transformers, because of their great numbers, contribute significantly more losses than substation transformers. In the late 1970s, the Bonneville Power Administration (BPA) requested that Battelle Pacific Northwest Laboratory survey and analyze major contributors to losses on Bonneville's 131 retail utility customers. Based on this survey, Battelle found that distribution trans-formers contributed 16.2 percent of the losses on investor-owned utili-ty transmission and distribution systems. They also discovered that substation transformers contributed only 4 percent of the losses on investor-owned utility transmission and distribution systems. They found that distribution transformer losses accounted for 36.5 percent of the losses and substation transformers only 2.2 percent of the losses on nongenerating public utility transmission and distribution systems. Figure 1-1 shows a pie chart of the components contributing to the losses on investor-owned transmission and distribution systems. Figure 1-2 shows a pie chart of the components contributing to the

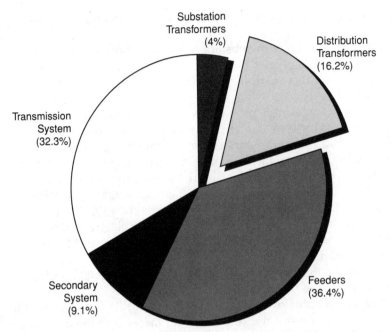

Figure 1-1. Investor-owned utility losses. (*Battelle Pacific Northwest Laboratory.*)

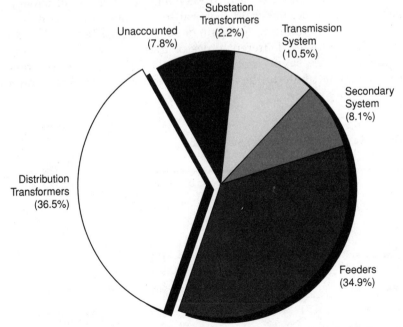

Figure 1-2. Nongenerating public utility losses. (*Battelle Pacific Northwest Laboratory.*)

losses of nongenerating public utility transmission and distribution systems.

This book deals with both substation and distribution transformers. However, due to the greater potential for loss savings with distribution transformers, this book focuses on energy-efficient distribution transformers.

Distribution transformers primarily come in two types: liquid-filled (usually oil) and dry-type. Electric utilities purchase and own 90 percent of the liquid-filled transformers, while all dry-type transformers are owned by commercial and industrial users of transformers. According to an Oak Ridge National Laboratory report entitled, *Determination Analysis of Energy Conservation Standards for Distribution Transformers,* "In the United States, approximately 40 million liquid-immersed distribution transformers are owned by electric utilities and an additional 4 million liquid-immersed units are non-utility-owned. Transformer manufacturers estimate that approximately 12 million dry-type distribution transformers are used by commercial and industrial customers in the United States."

Based on an Electric Power Research Institute (EPRI) study of a typical utility system, transformer core (no-load) losses amounted to about 1.4 percent of the electricity generated on a 5000-MW utility system, while transformer coil (load) losses amounted to about 1.3 percent of the electricity generated on a utility system. To increase efficiency, transformer manufacturers add material to the transformer core and coil with a corresponding weight gain or substitute core and coil material with lower losses. Different types of transformers have varying amounts of energy saving potential. Table 1-1 shows how different transformers contribute to the annual transformer energy losses on a 5000-MW utility system.

Table 1-1. Transformer Losses on a Typical Utility System.

Transformer type	Millions of kWh	
	Core	Coil
Generator stepup	18	89
Bulk power substation	67	138
Distribution substation	97	114
Distribution	328	127
Total	510	468

SOURCE: Courtesy of EPRI.

Energy Saving Potential

Today, very high efficiency transformers use specially manufactured steel, such as amorphous metal and laser-etched steel. Amorphous metal distribution transformers are available commercially. Replacement of the nation's existing 20 to 40 million distribution transformers with amorphous metal transformers could result in an estimated energy savings of 6 to 14 billion kWh annually. This could be worth between $300 to $700 billion a year. Even though it is unlikely that amorphous technology will ever completely supersede the use of conventional transformers, its introduction is resulting in significant changes in the transformer market. If you wish to learn more about low-loss transformers, turn to Chap. 8. And if you wish to learn more about amorphous transformers, go to Chap. 9 and read about the economical and technical status of amorphous transformers. A computer program showing how to evaluate the cost-effectiveness of amorphous metal and silicon steel transformers is included on a floppy disk in the back of this book.

The same report by the Oak Ridge National Laboratory entitled, *Determination Analysis of Energy Conservation Standards for Distribution Transformers*, estimates that there is a potential energy savings of 5.2 to 13.7 quads (quad = 1 quadrillion or 10^{15} Btu) over the 30-year period from 2000 to 2030. Figure 1-3 shows this potential savings graphically.

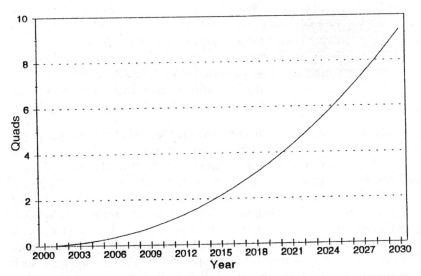

Figure 1-3. Cumulative potential transformer energy savings from 2000 to 2030. (*Oak Ridge National Laboratory.*)

With the potential for large energy savings from an overall perspective, how can you, the transformer purchaser, determine whether it is cost-effective to purchase an energy-efficient transformer? There are several methods for evaluating the cost-effectiveness of energy-efficient transformers. All these methods require comparing the cost of the transformer with the cost of the transformer losses. This can be done by comparing the cost of the transformer with the benefit of the loss savings from using energy-efficient transformers. Or it can be done by comparing the cost of various transformer designs and their corresponding losses. In all these cases, the transformer purchaser must know the cost of the transformer, the amount of the losses, the value of losses, and how to evaluate the losses over time.

The most common practice and industry cornerstone for determining the cost-effectiveness of energy-efficient transformers is calculating total owning cost. The utility industry developed this concept in 1981 as an industry standard. The *total owning cost* (*TOC*) provides the transformer purchaser with a means to compare various transformer designs. The transformer that meets the transformer purchaser's technical specification with the lowest total owning cost becomes the most cost-effective transformer. The following formula describes how to calculate transformer total owning cost:

$$TOC = NLL \times A + LL \times B + C \qquad (1.1)$$

where TOC = total owning cost
NLL = no-load loss in watts (W)
A = capitalized cost per rated watt of NLL (A factor)
LL = load loss in watts (W) at the transformer's rated load
B = capitalized cost per rated watt of LL (B factor)
C = the initial cost of the transformer, including transportation, sales tax, and other costs to prepare it for service

The total-owning-cost formula provides the basis for organizing and using this book. The first part of this book provides the necessary background to understand the components contained in this formula: no-load and load losses, cost of the transformer, value of losses. Chapter 6, "Transformer Economics," explains how to use this formula and other evaluation methods. Subsequent chapters apply this formula to specially designed transformers, such as amorphous, low-loss, and K-factor transformers, while Chap. 11 deals with future trends in the deregulation of the utility industry and development of new technologies, such as superconducting transformers, and how they will affect the use of energy-efficient transformers. Figure 1-4 illustrates in flowchart form how the various chapters of this book build on one another.

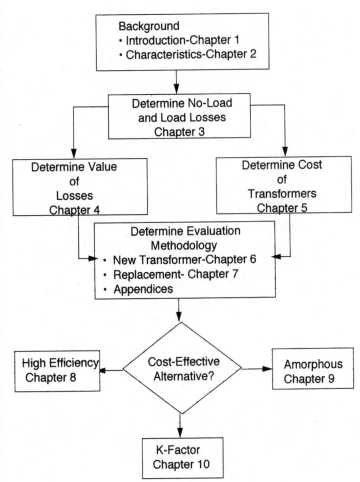

Figure 1-4. Energy-efficient transformer selection procedures.

With new technologies and existing technologies and the increasing cost of energy, the potential for energy conservation using energy-efficient transformers becomes more cost-effective. Using the techniques and knowledge described in this book will allow all users of transformers to pursue a conservation policy and save money at the same time. Chapter 2 will lay the framework for understanding transformers electrically and mechanically. If you are familiar with transformers already, you can skip Chap. 2 and go on to Chap. 3, "Transformer Efficiency."

2
Transformer Characteristics

The invention of the transformer had an impact on a controversy that waged in the nineteenth century. This major controversy's outcome would influence the use of electricity to this day. The controversy centered on whether electricity should be delivered at direct current (dc) or alternating current (ac). On the surface of the argument, the difference between dc and ac did not seem controversial. Dc delivers electricity at a constant voltage and current over time, while ac delivers electricity at a varying voltage and current over time, as shown in the two graphs in Fig. 2-1.

The great American inventor Thomas Edison promoted dc. By late 1887, he had built 121 central stations distributing dc power at 110 V that powered more than 300,000 of his incandescent lamps. Edison argued that dc was safer than ac. He even tested the safety of ac versus dc by electrocuting a horse. He seemed to ignore dc's inherent disadvantages.

Dc can only operate at the generator voltage. This is inefficient. This inefficiency of dc can best be understood by first knowing how voltage and current affect the efficiency of an electrical power system. Electrical energy is voltage times current. Thus, by raising the voltage and lowering the current, the same amount of electrical energy can be transmitted with less current. The magnitude of the current determines the size of the conductor. Therefore, a high-voltage line can transmit a large amount of power with a smaller current carried on a correspondingly smaller, less expensive conductor. Just as a bigger pipe is needed to transmit larger volumes of water, a bigger conductor

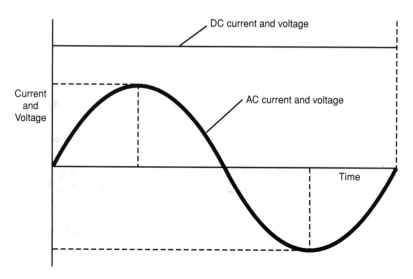

Figure 2-1. Current and voltage: ac versus dc.

is needed to transmit larger amounts of current. By increasing the pressure of the water in the pipe, you can increase the volume of water that can be transmitted without increasing the size of the pipe. The same thing is true of electricity. By increasing the voltage, you can increase the amount of transmitted power without increasing the size of the conductor. Also, losses in an electrical conductor are equal to the square of the current in the conductor multiplied by the resistance in the conductor. Losses reduce the efficiency of the conductor and waste energy and money. This concept is illustrated in the following formula:

$$W = I^2R \tag{2.1}$$

where W = power loss in watts (W)
 I = current in amperes (A)
 R = resistance in ohms (Ω)

George Westinghouse saw the advantages of ac over dc. He promoted the use of ac. He argued that ac had an economic advantage over dc. Ac allowed the generator voltage to be transformed to a higher voltage. The higher voltage would reduce losses and increase the amount of power the electrical system could transmit. Westinghouse knew that the transformer was essential to the economic advantage of ac over dc. He knew that the transformer's ability to increase voltage for economic transmission and lower voltage for safe use was essential to the economic success of ac. He knew that the transformer's function was based on the principle of electromagnetic induction, discovered in 1831 by the famous British physicist Michael Faraday.

Michael Faraday discovered that if he moved a magnet through a coil of copper wire, he caused an electric current to flow in the wire. Figure 2-2 illustrates what he discovered. Faraday found that voltage E_1, when applied to the primary coil, induced a corresponding voltage E_2 in the secondary coil with a magnitude that was directly proportional to the ratio of the turns in the two coils. He also discovered that the current I_1 flowing in the primary coil or conductor caused a corresponding current I_2 to flow in the secondary coil or adjacent conductor with a magnitude that was inversely proportional to the ratio of the turns (N_1/N_2) in the two coils or conductors. From this discovery he derived the following equation:

$$\frac{E_1}{E_2} = \frac{I_2}{I_1} = \frac{N_1}{N_2} \tag{2.2}$$

From this simple concept, the modern transformer was born. The transformer itself is an energy-conservation device. It transforms voltage and current. Consequently, it reduces losses in the transmission and distribution system. However, it does contribute some losses of its own.

The transformer is like a water pump. It raises the voltage so that more current can flow in the conductor with less loss, just as a water pump raises pressure so that more water can flow in the pipe with less friction loss.

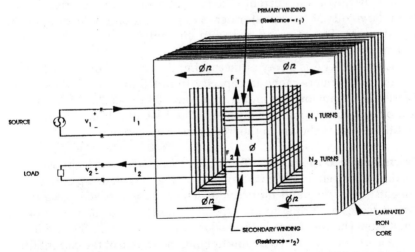

Figure 2-2. Magnetic effect in a transformer. (*Oak Ridge National Laboratory.*)

Transformer Components

Transformer components describe the transformer's characteristics and how they affect transformer losses. They can be divided into two categories: electrical and mechanical components.

There are several electrical and mechanical components that influence the energy consumption of the transformer. The electrical components include the kilovolt-ampere of kVA, the resistance, the voltage ratio, whether the transformer is three-phase or single-phase, and the basic insulation level or BIL. The mechanical components include the core, coils, tank, bushings, cooling system, and insulation medium. Let's first take a look at the electrical characteristics.

Electrical Characteristics

An understanding of the electrical characteristics of a transformer is critical to an understanding of how to use transformers efficiently. The first electrical component to consider is the kilovolt-ampere, a concept that describes the amount of energy the transformer will transmit and use.

Kilovolt-ampere (kVA). The *kilovolt-ampere* (kVA) rating of the transformer determines the physical size of the transformer. It also determines the size of the electrical load and load losses. It defines how the design engineer will design the transformer.

The transformer design engineer determines the transformer kVA by multiplying either the input voltage by the input current or the output voltage by the output current. The current flowing in the primary and secondary coils of the transformer is dependent on the loading of the transformer. The maximum design current flowing in the coils is dependent on the kVA of the transformer.

Thus using kVA is a way of measuring the capacity of a transformer. Power transformers range in capacity from 0.5 to about 3000 kVA for distribution transformers and from about 100 to 500,000 kVA for generation and substation transformers. The kVA of a transformer is the first thing that is shown on the transformer nameplate.

Transformer Impedance. Transformer *impedance* is the opposition of the flow of current in the transformer winding. It is made up of two components: resistance and inductive reactance. The *resistance* of a transformer varies according to the type of conductor and the length of the conductor. The resistance is the major cause of losses in the conductor. These losses are calculated by multiplying the square of the current times the resistance (I^2R). The resistance is a measure of the ability of the molecular structure of the conductor to transmit electrons. The magnitude of the primary winding resistance is calculated by multiplying the length of the conductor by the resistance per foot for the size of copper conductor used in the primary winding. The magnitude of the secondary winding resistance is calculated by multiplying the length of the conductor by the resistance per foot for the size of copper conductor used in the secondary winding. These two calculations are added to obtain the total resistance. The user of the transformer can determine the amount of resistance in a transformer from the transformer test report. Divide the load losses by the square of the rated current to get the total resistance of the transformer.

The second component of transformer impedance is the *inductive reactance*. The inductive reactance of the conductor causes the current to lag the voltage. Its value is determined by the characteristics of the winding and air gap between the primary and secondary windings. It has no effect on the losses of the transformer. The relationship between the inductive reactance, resistance, and impedance can be shown in the following formula:

$$Z = \sqrt{R^2 + X^2} \qquad (2.3)$$

where Z = impedance [Ω (ohms)]
$\quad R$ = resistance [Ω (ohms)]
$\quad X$ = inductive reactance [Ω (ohms)]

The impedance of a transformer is specified on the nameplate in percent. The percentage of impedance is determined by taking the ratio of impedance in ohms divided by the rated voltage of the transformer. The percentage of impedance can be converted to an ohm value by multiplying the percent impedance by the rated voltage.

Voltage Ratio.   The *voltage ratio* is the ratio of the number of turns in the secondary winding to the number of turns in the primary winding. The greater the turns ratio, the more turns there are in the conductor and a greater length there is of the conductor. This results in a larger resistance. Therefore, a larger turns ratio will result in more load losses.

Basic Insulation Level (BIL).   Transformers are designed to withstand system transits caused by switching or lightning surges. The transformer industry has set standards for transformers to withstand these various transits. These standards determine the amount of insulation required in the transformer. These standards of insulation are called the *basic insulation level* (BIL). The magnitude of BIL sets the distance between the primary and secondary windings and the amount of insulation on the conductor. The BIL rating affects the size and amount of losses that occur on a particular transformer. Table 2-1 describes the various BIL ratings developed by the transformer industry.

Three-Phase versus Single-Phase. Almost all transmission systems are three-phase. Consequently, any transformer designed to transform

Table 2-1. Basic Insulation Level (BIL)

Nominal system voltage (kV)	Basic impulse level (kV)	Low-frequency test to ground (kV)
1.2	45	10
2.5	60	15
5.0	75	19
8.7	95	26
15	110	34
25	150	50
34.5	200	70
46	250	95
69	350	140
115	550	230
115	450	185
115	350	140
138	650	275
138	550	230
138	450	185
161	750	325
161	650	275
161	550	230
230	1050	460
230	900	395
230	825	360
230	750	325
230	650	275
345	1300	575
345	1175	520
345	1050	460
345	900	395
500	1675	750
500	1550	690
500	1425	630
500	1300	575
725	2050	920
725	1925	860
725	1800	800

three-phase voltage is either one single three-phase transformer or three separate single-phase transformers. Distribution systems can be single-phase or three-phase. Three-phase transformer primary or secondary windings can be delta or wye connected. Figure 2-3 shows the relationship between line current and phase current in a three-phase delta-delta–connected transformer. In a three-phase delta-delta–connected transformer, the phase current equals the line current divided by the square root of 3.

Figure 2-4 shows the relationship between the line current and the phase current in a three-phase delta-wye–connected transformer. On the wye-connected transformer secondary, the phase current equals the line current. This is important to know when calculating total three-phase I^2R losses. When you are calculating losses in a three-phase transformer, it is necessary to multiply the I^2R losses in each phase by 3 to obtain the total losses.

Mechanical Characteristics

The physical or mechanical characteristics of a transformer provide important information on how to use transformers more efficiently.

DELTA - DELTA

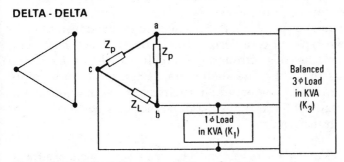

Figure 2-3. Delta-delta–connected transformer.

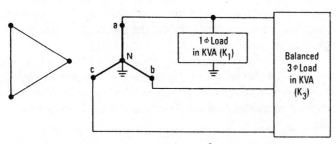

Figure 2-4. Delta-wye–connected transformer.

The ideal transformer is made of two major components: the core and the coils. The core acts as a transmitter of the electromagnetic lines of force. The coils act as a transmitter of the electric current in the transformer. A closer look at the core will show how it transmits electromagnetic lines of force.

Core.   The core can be constructed of air or iron. Air is a very inefficient means of transmitting the lines of electromagnetic force necessary to induce voltage and current in the coils. The iron core's molecular structure makes it a good conductor of electromagnetic lines of force or flux. Efficiency of the iron improves as the iron becomes softer. This is so because the molecules of the iron each become like little magnets when a magnetic field is applied to the iron. Softer iron takes less energy to magnetize and demagnetize. The ease by which the softer iron is magnetized influences the iron's ability to transmit the magnetic field.

The iron core experiences two types of losses: hysteresis and eddy current losses. These losses will be discussed in more detail in Chap. 3. The hysteresis losses are due to friction caused by the molecules resisting being magnetized and demagnetized. The eddy current losses are due to the electrical resistance in the core causing current to flow in the core in eddies like water flows in eddy pools in a stream. The core losses are sometimes referred to as *no-load losses,* because they occur in the core of the transformer when there is no load on the transformer.

There are two types of core design: core form and shell form. In a core-form design, the coils surround the core. In a *shell-form design,* the core surrounds the coils. The shell-form design tends to be more expensive to build but has better short-circuit strength than a core-form design. Figure 2-5 illustrates the difference between a core-form and shell-form transformer.

Coils.   The coils or conductors are the other main component of a transformer. The coils are divided into two parts: the primary coil and the secondary coil. The primary coil is where the source of electrical energy initially flows. The secondary coil is the coil that receives the induced transformed voltage and current. The coils are usually made of copper. The size of the coil conductor is determined by the size of the current and the desired magnitude of the losses.

The losses in the coils often are referred to as I^2R losses. This is so because the magnitude of the losses in the coil is equal to the current I multiplied by itself times the electrical resistance of the coil. This loss is sometimes referred to as the *load loss* because it varies according to the size of the load.

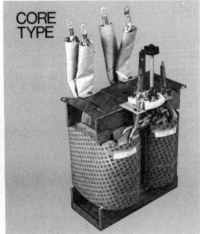

Figure 2-5. Shell-type and core-type designs. (*Courtesy of GE.*)

Tank. The transformer tank can be seen as a "black box." The transformer tank can be square or round in configuration. Its configuration
depends on the internal design of the core and coils of the transformer. It
comes in various shapes, sizes, and weights. Large generator and substation transformers are often rectangular in shape, while small distribution
transformers are cylindrical in shape. Distribution transformers can
weigh as little as 25 lb, while generation transformers can weigh as much
as 1 million lb.

The tank contains the core, coils, and insulating medium of the
transformer. It may contain braces to strengthen it against the large
forces that occur during a short circuit. It is often painted with a pig-
ment that dissipates the heat generated in the transformer. It is made
of a type of steel that reduces the transmission of the magnetic field.
Often it will be designed to contain the noise caused by the contraction
and expansion of the iron core due to the alternating magnetic field. It
may contain louvers or radiators to transfer heat to the surrounding
air. Because of the large amount of heat generated in the core and coils
of the transformer, it is necessary to design a cooling system in the
transformer to dissipate the heat to the surrounding air.

Cooling System. The losses in the transformer are dissipated in the
form of heat. Too much heat can be very detrimental to transformer operation. It is therefore important that the transformer be designed with a
cooling system to get rid of the heat. Often transformers will use radia-

tors to dissipate the heat, something like a car uses a radiator to keep the car running cool. Some transformers also use pumps to keep the cooling medium flowing. In the case of a liquid-filled transformer, the cooling medium is often oil. In the case of solid-state transformers, air in the transformer is used to keep it cool.

The cooling system consists of the insulating medium surrounding the core and coils. The insulating system can be some liquid, such as transformer oil, or air. The heat caused by the losses in the core and coils of the transformer affects the efficiency of the transformer. Rising temperature increases the resistance in the conductor. In addition, rising temperature will cause the insulation to deteriorate until the transformer fails. Consequently, the transformer design engineer designs the transformer to keep its temperature no more than 55°C above the ambient temperature. The designer will install louvers in large distribution transformers to radiate the heat and keep the transformer temperature within design limits. Large generator transformers may be forced oil water (FOW) transformers. The cooling system may consist of pumps to circulate the oil inside the transformer and pumps to circulate water outside the transformer. While many substation transformers have radiators through which the oil circulates by natural heat convection, some may have pumps to circulate the oil and fans to cool the oil circulating through the radiators. Figure 2-6 shows how the oil circulates in an oil-cooled transformer.

Often transformers operate in three stages, such as an OA/FA/FOA transformer. This type of transformer operates at a self-cooled rating, OA; FA, which is 133 percent of the OA rating with fans operating; and FOA, which is 167 percent of the OA rating with fans and pumps both operating. The fans and pumps necessary to cool some transformers use energy. This energy consumption thus reduces the efficiency of the transformer and must be included in any efficiency calculation. The fans and pumps operate at a low voltage (110 V) and often require a small distribution transformer to step the voltage down to this level. This small distribution transformer's energy consumption must be included in any calculation of the overall efficiency of the larger substation transformer. Some large substation transformers, in order to keep the height down, will have conservatory tanks on the top of the transformer to allow the oil to expand when it heats up. In any case, inside the top of the transformer tank is a cushion of nitrogen to allow for expansion of the oil during the transformer's operation. On the top of the outside of all transformer tanks are the bushings.

Bushings. Bushings are designed to connect the input and output of the transformer to the transmission and distribution system. They con-

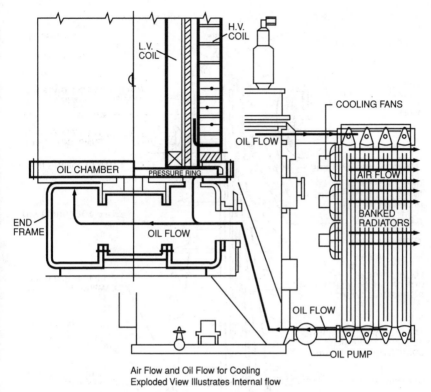

Air Flow and Oil Flow for Cooling
Exploded View Illustrates Internal flow

Figure 2-6. Oil-cooled transformer. (*Courtesy of ABB.*)

tain a conductor surrounded by a ceramic skirt. The skirt prevents the
voltage from creeping or arcing to the tank. The magnitude of the voltage affects the length of the bushings, while the magnitude of the current
affects the diameter of the conductor inside the bushing. The bushing has
an insignificant effect on the efficiency of the transformer. Figure 2-7
shows the construction of a bushing. Just as the bushing insulates the
tank from the outside terminals, an insulating medium is required inside
the tank to insulate the coils and terminals from the tank.

Insulation Medium.   There are basically two types of insulation
medium: dry or liquid. The dry-type transformer can be insulated by
either air or an epoxy. Dry-type transformers are usually small distribution transformers that do not generate much heat, while larger substation or distribution transformers require some type of liquid, such
as transformer oil, to dissipate the heat. Liquid-insulated transformers
that are in a confined area require a nonflammable insulation liquid.

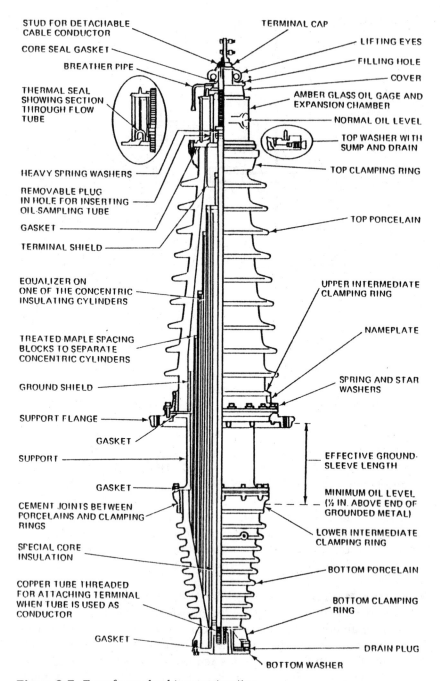

STUD FOR DETACHABLE CABLE CONDUCTOR

CORE SEAL GASKET

BREATHER PIPE

THERMAL SEAL SHOWING SECTION THROUGH FLOW TUBE

HEAVY SPRING WASHERS

REMOVABLE PLUG IN HOLE FOR INSERTING OIL SAMPLING TUBE

GASKET

TERMINAL SHIELD

EQUALIZER ON ONE OF THE CONCENTRIC INSULATING CYLINDERS

TREATED MAPLE SPACING BLOCKS TO SEPARATE CONCENTRIC CYLINDERS

GROUND SHIELD

SUPPORT FLANGE

GASKET

SUPPORT

GASKET

CEMENT JOINTS BETWEEN PORCELAINS AND CLAMPING RINGS

SPECIAL CORE INSULATION

COPPER TUBE THREADED FOR ATTACHING TERMINAL WHEN TUBE IS USED AS CONDUCTOR

GASKET

TERMINAL CAP

LIFTING EYES

FILLING HOLE

COVER

AMBER GLASS OIL GAGE AND EXPANSION CHAMBER

NORMAL OIL LEVEL

TOP WASHER WITH SUMP AND DRAIN

TOP CLAMPING RING

TOP PORCELAIN

UPPER INTERMEDIATE CLAMPING RING

NAMEPLATE

SPRING AND STAR WASHERS

EFFECTIVE GROUND-SLEEVE LENGTH

MINIMUM OIL LEVEL (½ IN. ABOVE END OF GROUNDED METAL)

LOWER INTERMEDIATE CLAMPING RING

BOTTOM PORCELAIN

BOTTOM CLAMPING RING

DRAIN PLUG

BOTTOM WASHER

Figure 2-7. Transformer bushing construction.

Polychlorinated bipthenyl (PCB) transformers were meant to fill this need. However, PCBs in transformers were found to be toxic to the environment. Thus most PCB transformers have been scrapped. No manufacturer makes PCB transformers anymore. Transformer manufacturers use other types of nonflammable liquids for transformers that are installed in confined spaces where a potential fire can be a safety hazard. The location and use of the transformer on the electrical system affects its design and energy consumption.

Transformer Locations

The main use of transformers is to change the voltage to an acceptable and useful level, although transformers are used for other purposes. For example, they are used to isolate one system from another; they are used to change the voltage and the current to allow the operation of instruments; and they are used extensively in the electronics industry. This book will be restricted to power transformers and their uses.

Power transformers are used to step the voltage up at generators. These are large transformers. The low-voltage winding usually operates at 13.8 kV, and the high-voltage winding is the operating voltage of the transmission system. In this book, generator transformers will be included with substation transformers. Substation transformers are usually used to step voltage down from the transmission system (greater than 115 kV) to the distribution level (69 kV or less). Figure 2-8 illustrates the various uses of transformers from the generator transformer to the pole-type distribution transformer that hangs on a pole just outside a residential area.

Transformers operate continually. They operate even when no electricity is being used. They are the silent economizers of the electrical power system, always making sure that the electrical power system is operating at the most economic voltage. They operate when everyone is asleep. Most of the time transformers operate without any human control. Sometimes an operator changes the voltage ratio using a no-load tap changer to regulate the voltage.

There are basically three locations and types of transformers: generation, substation, and distribution. We will examine each one of these types of transformers and where they are used.

Generation Transformer

The generation transformer is usually a large two-winding transformer located next to a generation facility, such as a hydroelectric

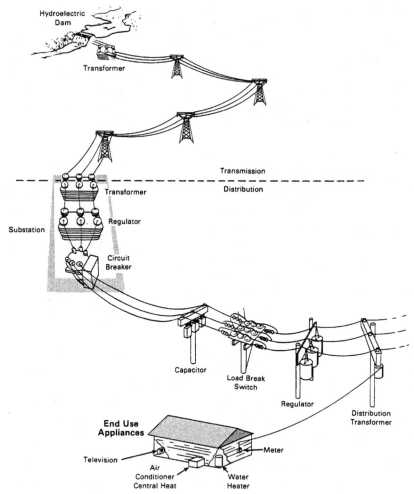

Figure 2-8. Transformer locations. (*Courtesy of Bonneville Power Administration.*)

dam, fossil fuel generator, nuclear generator, or gas-fired turbine generator. It is usually designed to step the voltage up from the output of the generator, e.g., 13.8 kV, to the high-voltage transmission level, e.g., 500 kV. Because of their sizes, these types of transformers often are built as three single-phase units to accommodate their transport to the installation site.

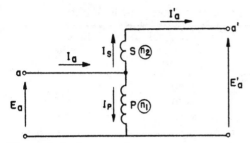

Figure 2-9. Autotransformer.

Substation Transformer

Substation transformers are located in substations usually to step the voltage down from an extra-high-voltage level, e.g., 500 kV, to a transmission level, e.g., 230 or 115 kV. They can be either three-winding or autotransformers. Three-winding transformers include a primary and secondary winding plus a tertiary winding. The tertiary winding is built to suppress third harmonics in shell-form transformers and to provide station service for the substation. They are usually three-phase transformers.

In an autotransformer, a single winding provides both the primary and secondary windings, according to the diagram in Fig. 2-9. The autotransformer results in a more efficient and smaller transformer with the same kVA capacity of a standard two-winding transformer. Autotransformers are not viable if the turns ratio exceeds a certain amount (usually a 3:2 ratio).

Distribution Transformer

The distribution transformer is designed to step the voltage down from a distribution level for use at an industrial, commercial, or residential facility. The type of facility determines the voltage level. Most industrial facilities operate at 480 V. The voltage coming into a house, however, is usually 220 V. The high side of the distribution transformer can vary from 13.8 to 69 kV. Table 2-2 lists examples of existing distribution transformers, their efficiencies, and their sizes.

Distribution transformers number in the millions and have the greatest potential for energy savings because of their vast numbers. They are located mounted on distribution poles, outside on a pad, or underground in a vault.

Table 2-2. Distribution Transformer Sizes and Efficiencies

Transformer size and type	2-year payback efficiency* (percent)	Recent purchase efficiency* (percent)
10-kVA pole, single-phase	98.40	98.33
15-kVA pole, single-phase	98.50	98.43
25-kVA pole, single-phase	98.70	98.67
37.5-kVA pole, single-phase	98.80	Not available
50-kVA pole, single-phase	98.90	Not available
50-kVA pad, single-phase	98.90	98.66
75-kVA pad, single-phase	98.00	98.83
167-kVA pad, single-phase	98.20	98.15
225-kVA pad, three-phase	98.00	98.84
500-kVA pad, single-phase	98.20	98.07
1000-kVA pad, single-phase	98.30	98.27

*All at 50 percent effective capacity factor.
SOURCE: Courtesy of ORNL.

Transformer Energy Use

Now that you have a good idea how transformers work and consume energy, let's look at the value of using energy-efficient transformers. Basically, the issue comes down to weighing the value of efficiency improvements against the increased capital cost of energy-efficient transformers. Using energy-efficient transformers has the benefit of reducing energy consumption and thus reducing the need to operate generators that dump heat and carbon dioxide into the atmosphere. If the electrical utility uses energy-efficient transformers, it reduces its operating cost and its need to raise rates. If the commercial or industrial owners of transformers use energy-efficient transformers, they benefit from lower power bills. With energy-efficient transformers, the utility experiences increased savings in load losses when the load increases. It also experiences savings in no-load losses even when the transformer is not loaded. Unlike other conservation measures, the utility experiences no loss in revenue with energy-efficient transformers.

In most cases, it is cost-effective to use energy-efficient transformers. In order to determine the cost-effectiveness of high-efficiency transformers, you must put a value to losses and evaluate losses as well as the capital cost of buying the transformer. How cost-effective are energy-efficient transformers? What efficiency rating is cost-effective? The next chapter examines the efficiency of various types of transformers and how their efficiency can be improved.

3
Transformer Efficiency

Substation and distribution transformers are very efficient devices. Their efficiency varies between 98 and over 99 percent. Thus, why is their efficiency a concern? This was the case for most users of transformers prior to the 1970s. They did not concern themselves with transformer efficiency. The oil crisis in the 1970s and the increased cost to produce electricity changed this attitude. The efficiency of transformers became a major concern. Reflecting this interest in transformer efficiency in 1981, the utility industry developed guidelines for evaluating losses on transformers. Chapter 8, "Transformer Economics," discusses these guidelines and other methods for evaluating transformer losses.

The large number of transformers result in a significant amount of lost energy. A recent study by the Oak Ridge National Laboratories found that liquid-filled distribution transformers owned by utilities alone contributed "61 billion kWh of annual energy," while dry-type distribution transformers owned by commercial and industrial users of electricity contributed an additional "60 to 80 billion kWh of electric energy on the customer side of the electric meter." Since the 1970s, most electrical utilities have been evaluating losses on transformers, although nonutility users of transformers continue not to evaluate losses. Consequently, transformer manufacturers have increasingly improved the efficiency of transformers purchased by utilities. This has not been the case with nonutility purchasers of transformers. Most nonutility users do not evaluate losses. Most utilities use liquid-filled outdoor distribution transformers, while industrial and commercial

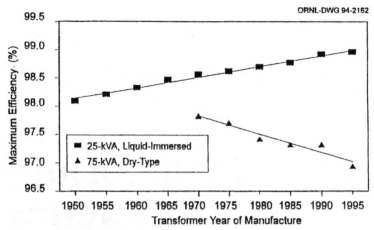

Figure 3-1. Maximum distribution transformer efficiencies versus time for a 25-kVA liquid-type and a 75-kVA dry-type distribution transformer. (*ORNL report.*)

users of transformers use indoor dry-type transformers. Figure 3-1 illustrates how manufacturers have responded to the increasing interest of utilities in improving the efficiency of transformers while building inefficient dry-type transformers for nonutility users of transformers.

Low-efficiency transformers affect the environment by requiring increased generation to supply the increased transformer losses. This wasted generation causes increased carbon dioxide in the air and contributes to the so-called greenhouse effect. Alternatively, energy-efficient transformers reduce the required generation and reduce the emissions into the air. A recent study by the U.S. Department of Energy concluded that utilities now replace a million transformers per year out of a total stock of 40 million transformers. If utilities replaced distribution transformers with higher-efficient transformers, they would reduce greenhouse gas emissions from the projected year 2000 levels by 0.8 million metric tons (MMT) of carbon equivalent.

In fact, Congress included a section on setting conservation standards for distribution transformers in the 1992 Energy Policy Act. This has resulted in both the U.S. Department of Energy and the Environmental Protection Agency studying the feasibility and significance of energy conservation for distribution transformers. Their studies were scheduled for completion in 1996, and these agencies will be developing standards for high-efficiency star transformers. But what is transformer efficiency?

What Is Transformer Efficiency?

The efficiency of any electrical machine is defined as the ratio of power input to power output. Electrical power is measured in watts (W) or kilowatts (kW), which is 1000 watts. The same is true of transformers, although a transformer's power rating is measured in kilovoltamperes (kVA) rather than kilowatts. Chapter 2 mentioned that the kVA rating defines the size and capacity of transformers. The transformer nameplate indicates the kVA rating. In order to determine the efficiency, the transformer kVA rating needs to be converted to kilowatts. How to convert kilovoltamperes to kilowatts involves the following simple formula:

$$\text{Power rating of transformers in kW} = (\text{kVA})(\cos \phi) \qquad (3.1)$$

where $\cos \phi$ = power factor. The power triangle shown in Fig. 3-2 describes this relationship between kilowatts and kilovoltamperes. Figure 3-2 shows that the $\cos \phi$ or power factor = kW/kVA. Therefore, for transformers, the efficiency can be defined by the following formula:

$$\text{Efficiency} = \frac{\text{Input kVA}}{\text{Output kVA}} = \frac{S \cos \phi}{S \cos \phi + \text{losses}} \qquad (3.2)$$

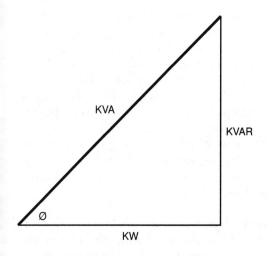

POWER FACTOR = COSØ = $\dfrac{\text{KW}}{\text{KVA}}$

Figure 3-2. Power triangle.

where S = kVA load
  losses = no-load losses + load losses $\times (S/S_B)^2$
  S_B = transformer kVA nameplate rating

If power factor or cos ϕ is unity, then the transformer efficiency formula becomes

$$\text{Efficiency} = \frac{S}{S + \text{no-load losses} + \text{load losses} \times (S/S_B)^2} \quad (3.3)$$

As can be seen from this formula, transformer efficiency is improved by simply reducing the transformer losses. Let's take a look at transformer losses, what causes them, and how to reduce them.

What Are Transformer Losses?

Total transformer losses can be divided into two components: no-load losses and load losses. The *no-load losses* occur in the core of the transformer at all times regardless of load, hence the term *no-load losses*, while the *load losses* occur in the transformer winding only when the transformer is loaded and vary according to the square of the load.

What Are No-Load Losses?

No-load losses occur in the transformer core 24 hours a day 365 days a year when a voltage is applied to the transformer regardless of the loading on the transformer. They are constant and occur even when the transformer's secondary is open-circuited. The no-load losses can be categorized into five components: hysteresis losses in the core laminations; eddy current losses in the core laminations; I^2R losses due to no-load current; stray eddy current losses in core clamps, bolts, and other core components; and dielectric losses. Hysteresis losses and eddy current losses contribute over 99 percent of the no-load losses, while stray eddy current, dielectric losses, and I^2R losses due to no-load current are small and consequently often neglected. The biggest contributor to no-load losses is hysteresis losses.

What Are Hysteresis Losses? *Hysteresis losses* are losses in the core laminations caused by the molecules resisting being magnetized and demagnetized by an alternating magnetic field. This resistance by the molecules causes friction that results in heat. The term *hysteresis* comes from a Greek word that means "to lag" and refers to the fact that the

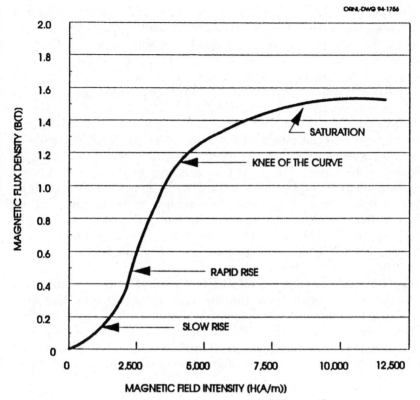

ORNL-DWG 94-1756

Figure 3-3. Typical initial magnetization curve. (*ORNL report.*)

magnetic flux lags behind the magnetic force. Figure 3-3 shows how this occurs. This is a graph of flux density versus magnetizing force. It is based on the following equation:

$$B = \mu H \qquad (3.4)$$

where B = flux density in the iron
μ = permeability of the iron
H = magnetizing force in the iron

As can be seen from this formula and Fig. 3-3, the hysteresis loop represents an increasing and decreasing magnetizing force that does not retrace its original track. This is so because the magnetic flux B, or lines of force, lag behind the magnetizing force H. The hysteresis loop illustrates graphically the nonlinear relationship between the magnetic flux density and magnetic field intensity. When applying a magnetic field to the ferromagnetic material in a transformer, the field should be kept at a level just below the knee in the magnetization curve. This is to avoid saturating the core and causing increased losses and harmonics. Keeping the peak operating flux from saturating the core results in a larger core and increased transformer capital cost, weight, and volume. The magnetizing and demagnetizing of the iron core cause hysteresis losses.

Hysteresis losses can be understood by examining Fig. 3-4, which illustrates the effect of an alternating magnetic field applied to the iron core of a transformer. The first diagram shows that each molecule of the iron becomes a tiny magnet when a magnetic field is applied to it. When the alternating field returns to zero, the molecules retain some residual magnetism. This residual magnetism resists realignment of the molecules when the magnetic field returns to its maximum value. The resistance of the molecules to being remagnetized causes friction and heat. This friction and heat are referred to as *hysteresis losses*. They amount to 50 to 80 percent of the total no-load or core losses.

Reduction of hysteresis loss is accomplished by changing the amount or type of iron. Increasing the amount of iron results in a reduction of the peak operating flux and hysteresis loss. This also causes an increase in transformer capital cost, weight, and volume. Silicon steel cores also reduce the amount of hysteresis loss by reducing the resistance of the molecules to being remagnetized. Amorphous steel cores result in a random configuration of the molecules that caus-

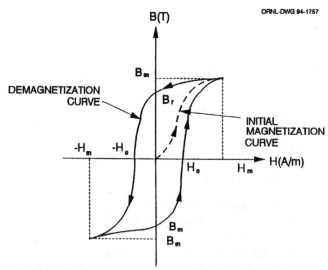

Figure 3-4. Typical hysteresis curve for ferromagnetic material. (*ORNL report.*)

es a significant reduction in hysteresis loss. Chapter 9 discusses more thoroughly amorphous metal transformers and how they reduce hysteresis loss. Another major contributor to losses in the core of a transformer is eddy current losses.

What Are Eddy Current Losses? *Eddy current losses* are losses caused by the current induced in the iron by the alternating magnetic field. As the magnetic flux changes with the alternating current in the coil, a current is induced in the iron that flows at right angles or cross-sectionally to the magnetic flux. This current is like the eddy currents that flow in a pool of water. The eddy current flow in the iron core causes I^2R losses that produce increased heat. The increased heat can result in increasing the resistance and consequently the losses. The eddy current losses contribute 20 to 50 percent of the total no-load losses. Figure 3-5 illustrates eddy current losses.

Building the core of thin laminated sheets of 7- to 12-mil thickness insulated from each other with a varnish made of a thin layer of insulating oxide reduces the eddy current losses. The thin laminations reduce the magnitude of eddy currents, while insulating each lamination breaks the flow of eddy currents in the core. Insulating the core laminations results in a more efficient transformer but increases the transformer capital cost, weight, and volume. In addition to eddy current losses, stray and dielectric losses occur in the transformer core.

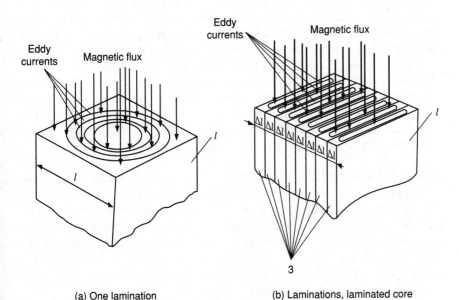

(a) One lamination   (b) Laminations, laminated core

Figure 3-5. Eddy currents in the core.

Stray losses are small and difficult to determine. They are caused by current being induced in clamps and bolts used in the installation of the core. They can be reduced by designing and manufacturing core clamps to minimize burrs. Steel manufacturers provide curves for determining stray losses.

Stray eddy current losses in the tank constitute another core-produced loss that is difficult to determine. It is caused by stray flux in the tank structure. It is reduced by using nonmagnetic material near bushing flanges and in the tank.

Even when there is no load on the transformer, there is an I^2R loss in the primary winding caused by the exciting current. The exciting current is the current necessary to magnetize or "excite" the core. This current flows in the primary winding and is about 1 to 5 percent of the load current. In larger transformers it is usually less than 2 percent of the load current. The losses associated with the excitation current are usually small and are neglected. They are reduced by reducing the resistance in the primary. This is accomplished by maximizing the cross-sectional area of the conductors and minimizing the length of the conductors.

What Are Load Losses?

Load losses are losses that vary according to the loading on the transformer. They consist of heat losses in the conductor caused by the load current and eddy currents in the conductor. These losses increase as the temperature increases. This is so because the resistance in the conductor is increased with temperature. It is often difficult to determine load losses because of the difficulty of knowing the load. This requires knowing the peak load as well as the load factor. The load factor can then be converted to the loss factor. Before describing how to determine the loading on a transformer, let's examine in more detail the types of load losses found in both substation and distribution transformers. The most significant load losses are I^2R or copper losses.

I^2R or Copper Losses. The I^2R losses are often referred to as *copper losses* because they occur in the copper windings of the transformer. They occur in both the primary and secondary windings. They are caused by the resistance of the copper conductor to the flow of current or electrons in the conductor. These losses are caused by the electrons moving in the conductor. The electron motion causes the conductor molecules to move and produce friction and heat. The energy generated by this motion can be calculated using the formula

$$kVA = (volts)(amperes) = VI \qquad (3.5)$$

According to Ohm's law, $V = RI$, or the voltage drop across a resistor equals the amount of resistance in the resistor multiplied by the current flowing in the resistor. Therefore, the kVA losses associated with the resistance in the conductor is calculated by substituting $V = RI$ in Eq. 3.5:

$$kVA = (RI)(I) \quad \text{or} \quad I^2R \qquad (3.6)$$

This is a significant loss and is equal to or greater than the no-load loss. How does the transformer designer reduce I^2R losses and make the transformer operate more efficiently?

The I, or current, portion of the I^2R losses cannot be changed by changing the design of the transformer. Only the resistance, or R, part of the I^2R can be changed. The resistance is reduced by using a material that has a low resistance per cross-sectional area without adding significantly to the cost of the transformer. Copper has been found to be the best conductor in terms of weight, size, cost, and resistance. Other than changing the conductor material, there are only two ways a transformer designer can reduce I^2R losses. One way is by increasing the cross-sectional area of the conductor. The other is by reducing the length of the conductor. A larger cross-sectional area allows the current to flow through the conductor with less motion of the molecules. The resistance of the conductor is directly proportional to the length of the conductor. Therefore, reducing the length of the conductor reduces its resistance. Another major contributor to load losses besides I^2R losses is eddy current losses in the conductor.

Conductor Eddy Current Losses.   Eddy currents form in the conductor of the transformer just like they form in the core of the transformer. They are due to magnetic field flowing perpendicular to the conductor causing eddy currents to flow in the conductor itself. Figure 3-6 illustrates the flow of eddy currents in a conductor.

Eddy current losses are reduced by transposing conductors. This reduces the voltage across the conductor that causes the eddy currents to flow in the conductor. Figure 3-7 illustrates how a transposition is constructed in a conductor.

The transformer loading over time is needed in order to calculate the load losses. This is not always easy to determine and requires certain assumptions. Factors have been developed to facilitate the calculation of loads and losses on transformers. These factors are called the *load, loss,* and *responsibility factors.*

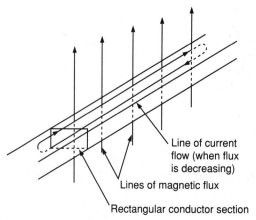

Figure 3-6. Eddy current in the conductor.

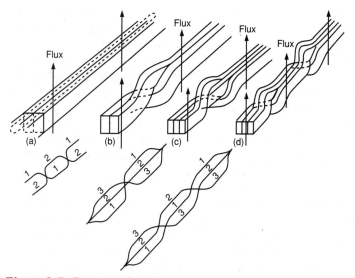

Figure 3-7. Transposed conductor.

Transformer Load, Loss, and Responsibility Factors

Transformer load factor is an important factor for determining load losses in power and distribution transformers. *Load factor* is defined as the ratio of average load in kilowatts to peak load in kilowatts, as shown in the following equation:

$$\text{Load factor} = \frac{\text{average load in kW}}{\text{peak load in kW}} \qquad (3.7)$$

The loss factor is determined from the load factor. The *loss factor* is the ratio of average losses in kilowatts to peak losses in kilowatts. The following formula gives a good approximation of how to derive the loss factor from the load factor:

$$\text{Loss factor} = 0.15 \times \text{load factor} + 0.85 \times \text{load factor}^2 \qquad (3.8)$$

Figure 3-8 shows this relationship graphically. The relationship between load and loss factor is dependent on the types of loads on the transmission and distribution system. It was derived from empirical measurements.

The loading on a particular transformer does not occur at the same time that the peak occurs on the transmission and distribution system. The *peak load responsibility factor* provides a method for determining what the peak loading would be on a specific transformer. The peak load responsibility factor is an estimate of the loading on a specific transformer at the time of peak loading on the transmission and distribution system. The *peak loss responsibility factor* is the square of the peak load responsibility factor. These factors allow you to calculate the loading on a transformer. In the case of a large generator or substation transformer, the peak load is usually determined by a power flow study. However, even with power flow studies, it is necessary to convert the peak transformer loading to annual transformer energy loading. This is where the load and loss factors can help by converting peak kilowatts to kilowatthours. The importance of these factors will become more apparent in Chap. 6, "Transformer Economics."

The no-load and load losses are common to all types of transformers, whether they are small distribution transformers or large substation transformers, while auxiliary losses normally are associated only with large substation transformers.

Auxiliary Losses

Auxiliary losses are losses caused by the use of cooling equipment such as fans and pumps to increase the loading capability of substation transformers. Figure 3-9 shows the fans and pumps in an OA/FA/FOA (self-cooled/fan-cooled/fan- and pump-cooled) transformer.

The energy consumption of auxiliary equipment depends on the horsepower of the fans and pumps and the length of time they are running. The length of time they are running depends on the transformer loading throughout the year. This can be determined from the

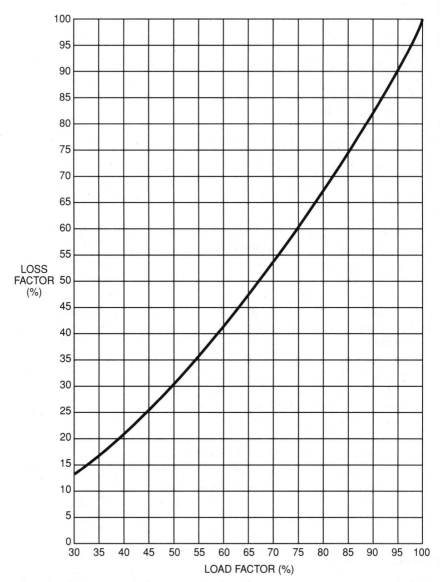

Figure 3-8. Load factor versus loss factor.

peak loading projections. Common practice is to turn on the fans when the transformer load reaches 133 percent of its rated load and to turn on the pumps when the transformer load reaches 167 percent of its nameplate rating. Some transformers are designed to run the fans and pumps continually. These are referred to as FOA transformers.

Figure 3-9. OA/FA/FOA transformer. (*Courtesy of GE.*)

In addition to the energy consumption of the fans and pumps, there is the loss associated with the distribution transformer necessary to step the voltage down to operate the fans and pumps. In a substation, the distribution transformer may be connected to the high-voltage feeder. A tertiary or third winding in the distribution transformer often is used to serve the load of the auxiliary equipment. The load loss of the tertiary winding needs to be included as part of the auxiliary loss-

Table 3-1. Auxiliary Equipment Energy
Consumption

Three-phase rating (MVA)	Auxiliary losses (KW)
5	1
10	1
15	2
20	2
25	3
30	4
40	4
50	4
60	5
80	6
100	7

SOURCE: Courtesy of Western Area Power Administration.

es. Table 3-1 provides some estimates of the energy consumption associated with various types of transformer auxiliary equipment.

How to Improve
Transformer Efficiency

Transformer efficiency is improved by reducing the losses in the transformer. The losses in the transformer can be reduced by changing the design, manufacture, operation, and maintenance of the transformer. The most common way to reduce losses is to change the design.

Transformer Design

The ideal design objective is to minimize the no-load losses and the load losses. If the no-load losses equal the load losses, the transformer has reached the ideal design to minimize losses. This is usually done by reducing core losses through a reduction in core flux density. However, reducing the core flux density to reduce core losses requires more conductor turns and increased load losses. Likewise, if the load density is lowered, more core material will be required, which will increase no-load losses. Transformer design to reduce losses is a compromise weighing the distribution of losses in the core and coils with the weight, size, volume, impedance, insulation, and cost of the transformer. Modern transformer design allows many variations in design variables through the use of computers in the design. Table 3-2 illus-

Table 3-2. Loss Reduction Alternatives

	No-load losses	Load losses	Cost
To decrease no-load losses			
A. Use lower-loss core materials	Lower	No change	Higher
B. Decrease flux density by			
1. Increasing core CSA*	Lower	Higher	Higher
2. Decreasing volts per turn	Lower	Higher	Higher
C. Decrease flux path length by	Lower	Higher	Lower
decreasing conductor CSA			
To decrease load losses			
A. Lower-loss conductor material	No change	Lower	Lower
B. Decrease current density by	Higher	Lower	Higher
increasing conductor CSA			
C. Decrease current path length by			
1. Decreasing core CSA	Higher	Lower	Lower
2. Increasing volts per turn	Higher	Lower	Lower

*CSA = cross-sectional area.

SOURCE: Courtesy of ORNL.

trates how changing core and conductor design can reduce no-load and load losses and improve transformer efficiency but also affect the cost of the transformer.

The impedance of a distribution transformer can affect the distribution and secondary system efficiency. In distribution transformers, too large of an impedance can cause voltage drops that result in lower voltage on secondary systems. Lower voltage on secondary systems causes higher currents and increased I^2R losses. Higher taps increase the secondary voltage but cause lower voltages on the primary side. The best way is to specify low impedances that meet the short-circuit needs of the system. Table 3-3 shows some impedance values for various sizes of distribution transformers.

Low impedance in substation transformers can result in increased system losses. Low-impedance substation transformers can cause more power to flow from the high-voltage system to the low-voltage system, resulting in higher system losses. For example, in an economic study it was found that it was cost-effective to replace a 30-year-old substation transformer with a new, more efficient transformer. However, the new transformer's impedance was 10 percent, while that of the old transformer was 15 percent. Placing the new transformer in the system and performing a power flow study showed that the savings in losses of the more efficient transformer were offset by the increased losses resulting from unloading the 230-kV system and loading the less efficient 115-kV system.

Table 3-3. Transformer Design Impedance

High-voltage rating (V)	Design impedance (percent)	
	Low voltage, rated 480 V	Low voltage rated 2400 V or higher
Power transformers		
2,400 to 22,900	5.75	5.5
26,400 to 34,400	6.0	6.0
42,800	6.5	6.5
67,000	—	7.0
Rated kVA	Design impedance (percent)	
Secondary-unit substation transformers		
112.5 through 225	Not less than 2	
Above 500	Not less than 4.5	
Network transformers		
1000 and smaller	5.0	
Above 1000	7.0	

SOURCE: IEEE Standard 241-1990.

Loss Distribution.   Loss distribution between no-load losses and load losses and between the primary load losses and secondary load losses affects the overall efficiency of the transformer. The highest efficiency occurs when the no-load losses equal the load losses and primary load losses equal the secondary load loss. This relationship can be shown by plotting the transformer efficiency curve as the ratio of the iron and copper losses is varied. It can be proven mathematically by taking the differential of the total losses and equating it to zero.

Transformer Manufacturing

Transformer manufacturing techniques that vary the transformer core configuration and materials continue to reduce no-load losses. Transformer manufacturers have attempted to maximize the reduction in no-load losses by the use of high-permeability, cold-rolled, grain-oriented silicon steel. They are seeking further reductions in no-load losses by using laser-etched steel.

Laser-etched steel causes the molecules of the steel to be oriented in the direction of the flux lines. This reduces the hysteresis losses by reducing the resistance of the steel from being magnetized and demagnetized. However, this method is costly and is only utilized when transformer users give high value to losses.

By matching steel laminations more accurately and using thinner laminations, transformer manufacturers have been able to greatly reduce eddy current losses. This maximizes the percent of core steel that bridges the gap at the core joint. This can be a costly and laborious process, since some large transformers require assembly of as many as 50,000 sheets and plates of steel in the core.

The use of amorphous metal in distribution transformers as large as 2500 kVA is commercially available. The molecular structure of amorphous metal core transformers is oriented in a random (amorphous) fashion rather than the orderly fashion of a silicon core transformer. When an alternating magnetic field is applied to an amorphous core, it takes less energy to magnetize and demagnetize the molecules than a silicon core transformer. This reduces the hysteresis losses significantly. Amorphous metal core transformers also use thinner laminations than silicon steel core transformers. The thinner laminations reduce the eddy current losses in the amorphous metal core transformer by reducing the resistance between laminations. The total reduction in no-load losses in an amorphous metal core transformer is 75 to 80 percent less than in a corresponding silicon steel core transformer. The amorphous metal core transformer saturates at a lower flux density than a silicon steel core transformer. Consequently, amorphous metal core transformers tend to be larger and weigh more than silicon metal core transformers of the same voltage and kilovoltampere rating. Chapter 9 discusses in more detail the loss savings characteristics of amorphous metal core transformers.

Auxiliary Equipment

Auxiliary equipment losses can be reduced by reducing the operating time of the auxiliary equipment. Such equipment is often controlled by thermocouples. Thermocouples inserted near the winding of the transformer detect the hot-spot temperature. When the hot-spot temperature is exceeded, the thermocouples send a signal to a transceiver to communicate to the fans and pumps to come on-line to keep the hot-spot temperature within tolerable limits. The hot-spot temperature is usually set not to exceed 65°C above the ambient temperature.

Transformer Sizing

Transformer sizing affects the efficiency of the transformer. The right sizing of transformers to fit the load keeps losses to a minimum. Oversizing a transformer can result in higher no-load losses, while undersizing a transformer can result in higher load losses. The right sizing of a transformer depends on the economic value and size of load losses versus no-load losses. This can be best understood by an example. The following is an example from the *Western Area Power Administration Distribution System Loss Evaluation Manual* (1988, pp. 5-6 to 5-7).

Example   This situation could be applied to an electric utility serving a new commercial customer or a commercial user of transformers. In this example, the peak load on the proposed three-phase transformer bank is expected to be 270 kVA with no known future increases. The transformer purchaser determines that a bank of three single-phase 75-kVA transformers will adequately carry the load, but three 100-kVA units also are being considered. Single-phase 75-kVA transformers from stock each have nameplate winding losses of 850 W at $1.33/W, core losses of 265 W, and a purchase price of $900. The single-phase 100-kVA units from stock each have nameplate winding losses of 1000 W, core losses of 350 W at $3.36/W, and a purchase price of $1100. Which transformer size would be more economical for the transformer purchaser?

solution   Although this is a three-phase installation, the characteristics on each phase are identical, so the transformer purchaser can perform the economic comparison on a single-phase basis using a peak loading of 90 kVA per phase. The comparison can be made by finding the losses and adjusted equivalent first cost for a single-phase 75-kVA unit and comparing them with the corresponding figures for a 100-kVA unit.

1. *75-kVA unit*

$$\text{Winding losses at 90 kVA} = (90 \text{ kVA}/75 \text{ kVA})^2 \times (850 \text{ W})$$
$$= 1224 \text{ W}$$

$$\text{Equivalent first cost of winding losses} = 1224 \text{ W} \times (\$1.33/\text{W}) = \$1628$$

$$\text{Equivalent first cost of core losses} = 265 \text{ W} \times \$3.36/\text{W} = \$890$$

$$\text{Total equivalent first cost} = \$900 + \$1628 + \$890 = \$3418$$

2. *100-kVA unit*

$$\text{Load losses at 90 kVA loading} = (90 \text{ kVA}/100 \text{ kVA})^2 \times 1000 \text{ W}$$
$$= 810 \text{ W}$$

$$\text{Equivalent first cost of winding losses} = 810 \text{ W} \times \$1.33/\text{W} = \$1077$$

$$\text{Equivalent first cost of core losses} = 350 \text{ W} \times \$3.36/\text{W} = \$1176$$

$$\text{Total equivalent first cost} = \$1100 + \$1077 + \$1176 = \$3353$$

These results show that the installation using 100-kVA transformers is the more economical choice. Although the evaluated cost difference is small, the 100-kVA units will provide extra capacity for unanticipated future load growth.

Losses are important but are not the only consideration in sizing transformers. Future replacement of an overloaded transformer needs to be considered as well.

Transformer Operation

Transformer operation can affect efficiency. Substation transformers with load tap changers can be operated in a way that minimizes the losses on the transmission and distribution system. By operating on the higher-voltage tap, the transmission and distribution system operates at a higher voltage and lower losses, although substation transformers with load tap changers have slightly higher load and no-load losses than substation transformers without load tap changers. Table 3-4 shows this difference.

An open-delta connection is often used to provide three-phase service from only two single-phase transformers. This connection causes the losses to be higher due to the transfer of kilovoltamperes from one transformer to another. This produces extra winding currents that do not serve the load. The extra losses may offset the savings of a third transformer.

Another way of saving losses is by operating transformers in a way that keeps the loading balanced. This is so because total losses are lowest when the load is divided among the transformers according to the nameplate rating. The following example from the *Western Area Power Administration Distribution System Loss Evaluation Manual* (1988; pp. 3-2 to 3-3) illustrates this principle.

Example A section of a utility's service territory is served by two substations, each of which includes a 10-MVA substation transformer with 50 kW of winding losses at rated load. Under peak conditions, substation A is loaded to 12 MVA and substation B is loaded to 6 MVA. What total winding losses occur on both transformers? If distribution line load transfers are made to equalize the loading on the transformers to 9 MVA each, what loss savings will result?

solution Load transfers will not affect no-load losses, and the effect on auxiliary losses will be minimal. Therefore, this analysis concerns only the load losses. Since load losses can vary with the square of the load, the peak load losses on each transformer can be calculated as follows:

$$\text{Winding losses} = (\text{MVA load}/\text{transformer size in MVA})^2 \times \text{rated load loss}$$

$$\text{Substation A winding losses} = (12\ \text{MVA}/10\ \text{MVA})^2 \times 50\ \text{kW} = 72\ \text{kW}$$

$$\text{Substation B winding losses} = (6\ \text{MVA}/10\ \text{MVA})^2 \times 50\ \text{kW} = 18\ \text{kW}$$

$$\text{Total winding losses} = 72\ \text{kW} + 18\ \text{kW} = 90\ \text{kW}$$

If the load is redistributed so that each transformer is loaded to 9 MVA, the losses on each transformer will be:

$$\text{Winding losses} = (9\ \text{MVA}/10\ \text{MVA})^2 \times 50\ \text{kW} = 40.5\ \text{kW}$$

$$\text{Total winding losses} = 40.5\ \text{kW} + 40.5\ \text{kW} = 90\ \text{kW}$$

$$\text{Losses saved} = 90\ \text{kW} - 81\ \text{kW} = 9\ \text{kW}$$

Table 3-4. Substation Transformers with and without Load Tap Changers

Three-phase rating (MVA)	No-load losses with tap changers (kW)	No-load losses without tap changers (kW)	Load losses) with tap changers (kW)	Load losses without tap changers (kW)	Annual difference in losses* (kWh)
5	10	9	31	30	11,000
10	17	16	53	50	17,000
15	23	22	70	67	16,000
20	29	27	86	82	28,000
25	34	32	101	96	31,000
30	40	38	115	109	33,000
40	50	47	141	134	45,000
50	60	57	165	157	48,000
60	70	66	188	179	59,000
80	87	84	232	221	73,000
100	106	101	272	259	78,000

*Annual kWh losses are based on transformer loading at nameplate, loss factor of 0.3, and auxiliary loss factor of 0.1.

SOURCE: Courtesy of WAPA.

Transformer Maintenance

Transformer maintenance involves inspection and testing, refurbishment, and retirements. Very few users of transformers maintain them to save energy. Maintenance is usually associated with reliability and safety. If a distribution transformer user can economically justify replacing a less efficient transformer with a more efficient transformer, a great deal of energy can be saved. In a study for the Department of Energy entitled, *The Feasibility of Replacing or Upgrading Utility Distribution Transformers During Routine Maintenance*, Oak Ridge National Laboratory found that "44 billion kWh annually could be saved if all distribution transformers were immediately replaced by new low-loss units." Chapter 6 presents the process for determining the cost-effectiveness of replacing or upgrading distribution transformers.

Transformer Efficiency Determination

Transformer efficiency is determined by measuring the no-load and load losses using two standard tests. Figure 3-10 shows the test setup for measuring load losses. In this setup, the low-voltage winding is short-circuited and the voltage applied to the high-voltage (primary) winding is adjusted until the full load current is read on the meter installed in the primary winding. The wattmeter will read the full load losses at rated current and load.

Figure 3-11 shows the test setup for measuring no-load losses in the transformer core. In this setup, the high-voltage (primary) winding is open-circuited. Rated voltage is applied to the primary winding, and the wattmeter on the primary side of the transformer measures the no-load or core loss.

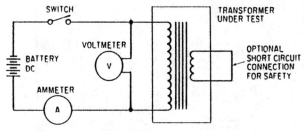

Figure 3-10. Load loss test setup.

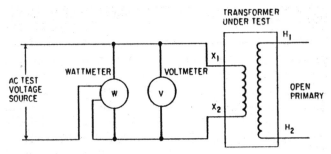

Figure 3-11. No-load loss test setup.

Test Data

Test data can be certified by the transformer manufacturer. In the case of large substation transformers, each transformer is tested for losses. In the case of distribution transformers, the manufacturer performs loss tests on a design and certifies that copies of that design have the same efficiency. The purchaser of transformers should perform random tests to verify the manufacturer's claimed efficiency ratings.

Nameplate

The transformer nameplate provides no information as to the efficiency of the transformer. As can be seen from Fig. 3-12, it provides basic information on the kilovoltampere rating, voltage ratio, impedance, and weight of the transformer.

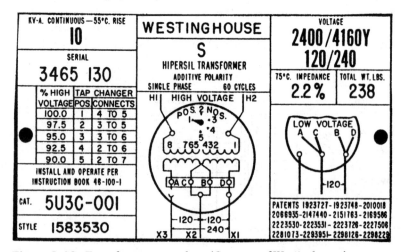

Figure 3-12. Transformer nameplate. (*Courtesy of Westinghouse.*)

Calculating Losses and Efficiency

In order to calculate the loss savings necessary to determine what transformer to buy based on an economic evaluation of the losses, it is necessary to determine the losses of the various transformers being considered. This can be done by comparing each transformer with a base case and then calculating the savings with the base case. However, if each transformer is compared on its own, then the absolute losses for each transformer are sufficient. It is important to keep the no-load losses and load losses separate when comparing transformers. Efficiency values are useful for rating transformers but are not usually provided when the transformer is purchased. The manufacturer instead usually provides no-load and load losses based on loss tests. Converting the losses into efficiency values becomes a simple straightforward calculation. The following example illustrates how to calculate efficiency.

$$\text{Efficiency} = \frac{\text{output}}{\text{input}} \times 100 = \frac{\text{output}}{\text{output} + \text{losses}} \times 100 \qquad (3.9)$$

The efficiency of a transformer is seldom specified directly, although this was the practice in the past. The present practice is to specify the no-load and total losses. This can be illustrated by an example.

Example The no-load loss of a 5000-kVA, 69,000-V transformer is 14 kW, and the total loss at full load is 41 kW. What is the efficiency at 0.5 load and at full load?

solution

$$\text{Total loss} = 41 \text{ kW}$$

$$\text{No-load loss} = 14 \text{ kW}$$

$$\text{Load loss at full load} = 41 - 14 = 27 \text{ kW}$$

Since the load loss varies as the square of the load, the load loss at 0.75 load is

$$27 \text{ kW} \times (0.5)^2 = 27 \text{ kW} \times 0.25 = 6.75 \text{ kW}$$

$$\text{Total loss at 0.5 load} = 14 \text{ kW} + 6.75 \text{ kW} = 20.75 \text{ kW}$$

$$\text{Total loss at full load} = 14 \text{ kW} + 27 \text{ kW} = 41 \text{ kW}$$

$$\text{Efficiency at 0.5 load} = \frac{2500 \times 100}{2500 + 20.75} = 99.18$$

$$\text{Efficiency at full load} = \frac{5000 \times 100}{5000 + 41} = 99.19$$

Determining No-Load Losses.   Determination of no-load losses is usually very simple and straightforward. No-load losses are the same 24 hours a day 365 days a year. They can be determined from the manufacturer's no-load loss measurements. They are dependent on the kilovoltampere size, voltage, and design of the subject transformer. They must be kept separate from the load losses because they have a different monetary value. They are usually worth more because they are constant and do not vary with time. Figure 3-13 shows how no-load losses remain constant in relationship to load losses.

Determining Load Losses.   Determining load losses is more complicated than determining no-load losses because they vary according to the load over time. This is illustrated in Fig. 3-13. Because load losses vary with the load, it is necessary to determine the loading of the system. This is usually given in terms of peak load. The peak load must be converted into loading over time. This conversion is accomplished by using the

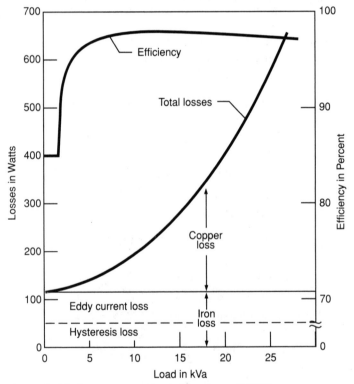

Figure 3-13. Losses versus load. (*Courtesy of EPRI.*)

load factor. Load factor's definition was given in Eq. 3.7. Once the load factor is determined, then the losses at the peak loading are used to determine the corresponding losses over time. The losses are determined by the loss factor using the following formula (Eq. 3.8):

$$\text{Loss factor} = 0.15 \times \text{load factor} + 0.85 \times \text{load factor}^2 \quad (3.8)$$

This formula is based on empirical results and may vary from system to system. Once the losses are determined for the peak loading of the system, then the losses for the corresponding loading on the transformer can be determined. This is done by using the capacity factor:

$$\text{Capacity factor } (CF) = \frac{\text{transformer peak load}}{\text{transformer rated load}} \quad (3.10)$$

The transformer peak load is not always easy to determine. It can be determined from a load flow analysis. Such an analysis is usually available for substation transformers but is not always available for distribution transformers. Often, with a distribution transformer, the transformer peak may occur at a different time from the feeder peak. The transformer peak can be found by using the responsibility factor:

$$\text{Responsibility factor} = \frac{\text{feeder peak load}}{\text{transformer peak load}} \quad (3.11)$$

Losses are measured at rated load. Therefore, in order to determine the transformer losses at transformer peak load, the capacity factor (Eq. 3.10) must be determined first.

Finally, the transformer load energy losses in kilowatthours can be determined:

$$\text{kWh} = \text{loss factor} \times CF^2 \times \text{losses in kW} \times 8760 \, \text{h/yr} \quad (3.12)$$

The flowchart in Fig. 3-14 illustrates how these factors come together in determining the annual losses on a transformer. The following example will illustrate how to use this formula in calculating total losses for a distribution transformer from the test total losses.

Example A liquid-type 25-kVA distribution transformer has 58 W of no-load losses and 312 W of load losses at rated load, a capacity factor of 1.2, and the loss factor is 0.3. What is the total energy loss in a year?

$$\text{kWh} = (\text{no-load losses} + \text{loss factor} \times CF^2 \times \text{load losses}) \times 8760 \, \text{h/yr}$$
$$= (58 + 0.3 \times 1.2^2 \times 312) \times 8760 = 1688 \, \text{kWh}$$

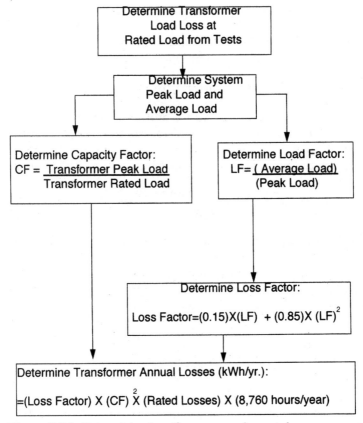

Figure 3-14. Determining transformer annual energy losses.

Transformer Efficiency Labeling

At present, there is no labeling of transformers for efficiency. Transformer efficiency must be obtained from the test report. This is unlike motors, which have their efficiency designated on the nameplate. Labeling of transformer efficiency would be a step that would encourage the use of more efficient transformers because it would provide the user with immediate information on the efficiency of new and used transformers. A first step for the industry to start labeling transformer efficiency is the development of transformer efficiency standards.

Transformer Efficiency Standards

Transformer efficiency standards do not exist at this time. The Department of Energy is in the process of analyzing the benefits of developing conservation standards for distribution transformers. This analysis is required by the Energy Policy Act of 1992. The Department of Energy analysis was done by Oak Ridge National Laboratories. The purpose of the analysis was to determine "whether energy conservation standards for distribution transformers would have the potential for significant energy savings, be technically feasible, and be economically justified from a national perspective."

The analysis found that it would be technically and economically feasible to develop conservation standards for distribution transformers. In fact, based on this analysis, Oak Ridge National Laboratories concluded that transformers account for an estimated 140 billion kWh of annual energy lost in the delivery of electricity and 2.51 to 10.7 quads (1 quad = 10^{15} British thermal units) of energy could be saved from the year 2004 to the year 2034. In addition to the effort by the Department of Energy to determine the feasibility of developing conservation standards for distribution transformers, the U.S. Environmental Protection Agency (EPA) has developed a voluntary program to encourage the use of energy-efficient transformers to reduce energy waste and greenhouse-producing carbon dioxide gases. The EPA calls this program the *Star Transformer Program.*

Star Transformers

The Star Transformer Program was introduced by the EPA in 1995. The EPA worked with the National Electrical Manufacturers Association (NEMA) to set transformer efficiency ratings that are feasible for transformer manufacturers to achieve. Utilities and manufacturers can voluntarily sign up to participate in the program. The energy star efficiency rating is achievable by both amorphous and silicon steel core transformers. Major manufacturers, such as ABB, General Electric, Cooper Power Systems, and Howard Industries, have already signed up to participate.

The participants in the EPA Star Transformer Program agree to purchase energy-efficient transformers that meet the star transformer efficiency ratings to the tune of 50 percent the first year, 70 percent the second year, and 95 percent thereafter. They agree to these targets by signing a memorandum of understanding with the EPA. They, of course, only purchase transformers that meet EPA's efficiency ratings and are cost-effective to buy. Chapter 6 explains how utilities determine the cost-effectiveness of purchasing energy-efficient transformers. Before getting into how to determine the cost-effectiveness of various transformer efficiency ratings, it is necessary to look at the various methods for determining the value of losses. How to value transformer losses is the subject of the next chapter.

4

Value of Losses

The value of transformer losses is important to the purchaser of a transformer. If the transformer purchaser assumes a high value for transformer losses, he or she will purchase more efficient transformers, or if the purchaser assumes a low value of losses, he or she will purchase less efficient transformers. But what value should the transformer purchaser assume? Thus far subjects dealing with energy-efficient transformers have been straightforward and factual. Now comes the murky and controversial subject of "values." It would be presumptuous to assume that this subject can be treated lightly without some controversy. Hopefully, it will not be as controversial as the present political debate about so-called family values. Before getting into the equations and ways to calculate the value of losses, it is worthwhile to discuss the thinking behind the determination of the value of losses.

Why the *value* of losses and not the *cost* of losses? This formulation is chosen because, ultimately, the determination of what losses are worth to the transformer purchaser is a value judgment. The choice may be to base that judgment on the cost of losses. Or it may be a value based on other factors that are difficult to quantify, such as conservation ethics, reducing pollution associated with energy production, or concerns about the diminishing worldwide energy resources. An electric utility may value losses differently than an industrial or commercial end-user of electricity. However, determining the value of transformer losses requires considering the rationale and assumptions for calculating the cost of losses.

How to determine the cost of losses may seem straightforward and simple at first glance. It is basically the cost to produce, transmit, and distribute each kilowatt of transformer loss. The electric utility must add capacity to its generation, transmission, and distribution system to deliver each additional kilowatt required to supply and deliver all losses, including transformer losses. In addition to the cost of generating, transmitting, and distributing capacity for transformer losses, there is the cost of generating, transmitting, and distributing the electrical energy. Both capacity and energy have to be dealt with individually. The utility faces some major problems in making this determination. What part of the generation, transmission, and distribution system should be counted for the cost of transformer losses? How does the cost of no-load losses differ from the cost of load losses? How does one account for the time value of money? How does one project the cost of losses into the future, since transformers have a lifetime average of 30 years? How does one handle inflation? How does the industrial and commercial end user deal with the uncertainty of electric utility costs and rates? How will retail wheeling affect the cost of losses? How will the breakup of utilities into transmission, distribution, and generation business lines affect the value of losses? How does the utility deal with a shrunken planning horizon of only 5 years or less? Should the transformer evaluation be based on levelized annual cost or capitalized cost? This chapter presents how to deal with these issues in a step-by-step procedure for calculating the value of losses.

First, it is important to recognize that the perspective of the electric utility is different from the perspective of the industrial and commercial users of transformers. The transformer loss evaluation procedure for the electric utility involves understanding and assessing the total cost of generating, transmitting, and distribution transformer losses. While the transformer loss evaluation procedure for industrial and commercial users requires an understanding and assessment of the electric rates they pay to the electric utility.

Electric Utility Perspective

The electric utility perspective for evaluating transformer losses is to evaluate the cost to generate, transmit, and distribute the power necessary to supply capacity and energy to the distribution transformer losses. Unlike the distribution transformer, no power flows through the distribution system to deliver substation transformer losses. Therefore, the electric utility does not include the cost to distribute the power in calculating the cost of substation transformer losses. This is so because the substation transformer loss is consumed before it reaches the distribution system. Figure 4-1 illustrates what parts of the utility's system contribute to distribution transformer losses and what parts contribute to substation transformer losses.

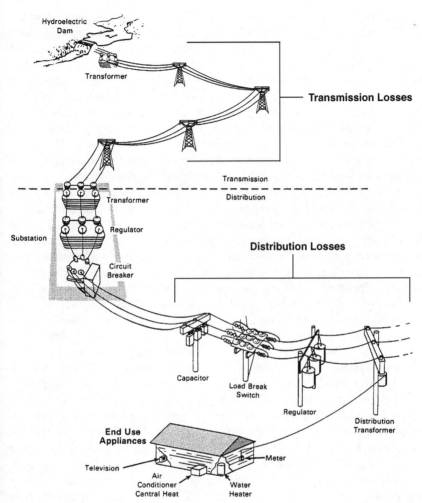

Figure 4-1. Parts of utility system contributing to transformer losses. (*Courtesy of Bonneville Power Administration.*)

If the utility does not generate its own power, then it must evaluate the cost of purchasing power from its power provider. Each electric utility may have its own method for evaluating transformer losses. Each method must have the basic components necessary to determine the cost of no-load losses separate from the cost of load losses. Besides the methods developed by individual utilities, there are methods developed by various organizations. These organizations have set methodology standards for the utility industry. These organizations include the Edison Electric Institute (which no longer develops standards for the utility industry), Institute of Electrical and Electronics

Engineers (IEEE), the former Rural Electrification Administration (REA), and the Department of Energy. Each one of these organizations has developed methods for evaluating transformers using the concept of total owning cost (*TOC*). The concept of total owning cost was introduced in Chap. 1. It allows the purchaser of transformers to compare transformers of different designs, costs, and losses on an equitable basis. When using the concept of total owning cost, it is necessary to decide whether to calculate capitalized total owning cost or annualized total owning cost.

Capitalized total owning cost involves converting the annual loss values into a capitalized value by dividing by the fixed carrying charge rate (see App. C). The fixed carrying charge rate is the levelized annual cost of the transformer divided by the cost of the transformer investment. Then the capitalized total owning cost for losses can be added to the bid price of the transformer to determine the capitalized total owning cost. *Levelized annual cost* refers to equal annual payments to pay off the capitalized cost of the transformer over the life of the transformer. The method developed by the IEEE utilizes the capitalized total owning cost method. The Department of Energy study by Oak Ridge National Laboratories also uses the capitalized total owning cost method. This method applies to both distribution and substation transformers. The only difference between the capitalized total owning cost method and the *annualized* total owning cost method is the role of the carrying charge rate. Multiply the bid price by the carrying charge rate in the capitalized total owning cost method, and divide the levelized annual cost of losses by the carrying charge rate in the annualized total owning cost method. Both methods are acceptable. According to REA Bulletin 61-16 (p. 8): "Although both the annual cost method and the equivalent first cost method are equally acceptable, the equivalent first cost is usually preferred." The method for calculating total owning cost for the substation transformer is different from the method for the distribution transformer.

Distribution Transformer

The distribution transformer evaluation method was described in Chap. 1. It is the foundation of the capitalized total owning cost method. In this formula, the total owning cost is the transformer first price levelized over the life of the transformer plus the cost of future transformer losses (i.e., no-load and load losses) discounted to present value and levelized over the life of the distribution transformer. To be consistent, in this chapter the capitalized total owning cost method will be used to evaluate the cost of distribution transformer losses. The costs of no-load and load losses are, respectively, referred to as the *A* and *B* factors in the total owning cost formula (Eq. 1.1):

$$TOC = NLL \times A + LL \times B + C \qquad (1.1)$$

In this formula, the two factors affected by the value of losses are the capitalized cost per rated watt of no-load loss or *A* factor and the capitalized cost per rated watt of load loss or *B* factor. Consequently, load and no-load losses have different values. Assignment of values to load and no-load losses will be dealt with separately.

Loss Multiplier (*L*).   Loss multiplier is a factor to account for the increase losses on the bulk transmission system resulting from transformer losses. Transformer losses become incrementally larger as a result of the increased bulk transmission losses. This is often called losses on losses. This quantity is usually measured in per unit and is a function of the bulk power system.

The value of the loss multiplier is equal to the ratio in percent of the transmission and distribution losses to the total generation. This value can vary from 1 to 20 percent depending on the efficiency of the bulk power system. For example, the efficiency of the Bonneville Power System is 1½ percent. This would result in increasing the losses on a distribution transformer by 1½ percent. In this case, a distribution transformer with 100 W of core loss would have 101 W of losses. In the case of a bulk power system with 10 percent losses, the core losses would increase to 110 W.

Value of Distribution Transformer No-Load Losses or *A* Factor.   The value of no-load losses, or the A factor, remains constant throughout the life of the transformer. Its value is determined by the capacity and energy required to generate, transmit, and distribute no-load transformer losses. It does not concern itself with the loading that may change daily on the transformer. Its value is the same 24 hours a day, 365 days a year, for 30 years. Because no-load losses are constant from a utility's perspective, the power to serve no-load losses comes from the utility's base load demand. It is concerned with the incremental cost of that base load demand. Even though the value of no-load losses presently varies throughout the United States, the trend toward deregulation seems to be causing the value to be the same from one utility to the next. The difficulty comes in assigning a value over the 30-year life of the transformer. The utility tries to project the value of no-load losses by predicting how the cost of generation and transmission and distribution will change in the future. The factors the utility needs to consider in the determination of the future change in the cost of losses will be discussed in the latter part of this chapter.

The initial utility method for determining the value of no-load losses was derived for distribution transformers. Even though the electric utility industry could adapt this method to apply to substation transformers, it lacked a definitive method for evaluating no-load and load losses on substation transformers. The Institute of Electrical and Electronics Engineers (IEEE) decided the electric utility industry needed a standard for evaluating the losses of substation transformers. The Transformers Committee of the IEEE Power Engineering Society was assigned to develop a standard for evaluating substation transformer no-load losses. The Transformers Committee developed in 1991 a method for evaluating substation transformer no-load and load losses entitled *IEEE Loss Evaluation Guide for Power Transformers and Reactors* (C57.120 91). This method is very similar to the method for distribution transformers, except that it deals with auxiliary power losses, whereas the distribution transformer method does not deal with auxiliary power losses. The distribution transformer method calculates the annualized total owning cost whereas the IEEE power transformer method uses the capitalized total owning cost method. In this book, to avoid confusion, assume that the capitalized total owning cost method is being used unless otherwise specified. Therefore, the cost of no-load losses or factor *A* is calculated according to the following formula:

$$A = \frac{SC + (EC \times 8760)}{FC} \times \frac{1}{1000} = \text{cost of no-load loss, in \$/W} \quad (4.1)$$

where *SC* = annual cost of system capacity in $/W-year. *SC* is the levelized annual cost of generation, transmission, and primary distribution capacity required to supply 1 W of load to the distribution transformer coincident with the peak load.

  EC = energy cost. *EC* is the levelized annual cost per kilowatthour of fuel, including inflation, escalation, and any other fuel-related components of operation or maintenance costs that are proportional to the energy output of the generating units.

  8760 = hours per year

  FC = fixed charge on capital per year. *FC* is the levelized annual revenue required to carry and repay the transformer investment obligation and pay related taxes, all expressed as a per unit quantity of the original.

  1/1000 = conversion from kilowatts to watts.

The process for solving this equation is illustrated in Fig. 4-2.

Before getting into an example of how to use this formula, it is important to understand each one of these terms and how to derive the value of each

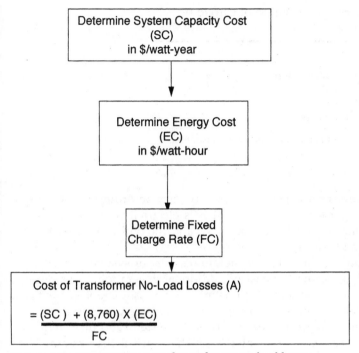

Figure 4-2. Determining cost of transformer no-load losses.

term. Each term can be best understood by explaining the components that make up that term. Even though the distribution transformer examples are 15 years old, the examples it contains for determining these terms are still illustrative today.

System Capacity Cost (SC). System capacity cost is the levelized annual cost of additional generation, transmission, and distribution capacity necessary to supply 1 kW of peak load to the distribution transformer. It does not include the cost of distribution capacity in the case of the substation transformer. It is the first component in calculating the cost of no-load and load losses. It reflects the cost of peaking generation and transmission and distribution capacity. It includes the cost of generation in base-load plants as well as the cost of generation in reserves. It varies from utility to utility. Table 4-1 illustrates the marginal value of capacity for the Bonneville Power Administration. The capacity cost in Table 4-1 does not include the cost of distribution capacity. This is so because the Bonneville Power Administration is a large bulk supplier of electrical power with very little distribution capacity.

Table 4-1. Marginal Value of Capacity
(1992 Dollars)

	Annual cost, $/kW
Levelized value of generation	41.92
Cost of transmission	13.90
Less transmission adjustment	−6.95
Total value of capacity	48.87

SOURCE: Courtesy of Bonneville Power Administration.

Typical values for generation capacity vary according to the type of fuel used for generation. The typical 1994 electricity costs for a gas turbine plant vary from $400 to $500/kW, and the typical 1994 costs for a fossil fuel plant vary from $1500 to $1600/kW.

In determining the value of the annual cost of system capacity, the utility needs to know the cost of additional generation, transmission, and distribution. One source of this information is the replacement cost values available from the Securities and Exchange Commission (SEC). First, determine the cost of each part of the system, and then divide by the present peak load associated with that part of the system. Then multiply each system investment per watt by the corresponding fixed-charge rate to levelize the cost of system capacity over the life of the transformer. The corresponding operation and maintenance rate is added to each component. The following example from the 1981 EEI report illustrates how to determine the system capacity cost:

Distribution SEC replacement cost	$300,000,000
Transmission SEC replacement cost	$240,000,000
Generation SEC replacement cost	$1,100,000,000
Distribution peak load	1300 MW
Transmission peak load	1700 MW
Generation peak load	1800 MW

Dividing each replacement cost by the corresponding peak loads yields the following results:

Distribution $/kW = $300,000,000 ÷ 1300 MW = 231

Transmission $/kW = $240,000,000 ÷ 1700 MW = 141

Generation $/kW = $1,100,000,000 ÷ 1800 MW = 611

Next, it is necessary to convert the investment cost per capacity (or $/kW) to an annual levelized value by multiplying by the fixed carrying charged rate for each respective component of the system. The fixed charge rate includes the rate of return, depreciation, income tax, dispersion allowance, and property taxes and insurance. In this example, the system capacity charge for distribution, transmission, and generation components of the system have the values listed in Table 4-2. The next common component to the value of no-load and load losses is the energy cost.

Energy Cost (EC). Energy costs include any cost proportional to the energy output of the generator. This would include the cost of fuel and the cost to operate and maintain the transportation, storage, and conversion of the fuel to electrical energy. For example, in the case of a fossil fuel plant, it would include the cost of the coal or natural gas and the cost to operate and maintain the equipment to transport the coal or natural gas and operate and maintain the equipment necessary to convert fossil fuels to electrical energy.

The assumed average fuel cost should be escalated over the useful life (n years) of the transformer and levelized by multiplying by the capital recovery factor (CRF_n; see App. G). An escalation factor (EF; see App. J) can be calculated using the following formula.

$$EF = \left[\frac{1 - \left(\dfrac{1+a}{1+i} \right)^n}{i - a} \right] CRF_n \qquad \text{for } a \neq i$$

and (4.2)

$$EF = \left(\frac{n}{1+i} \right) CRF_n \qquad \text{for } a = i$$

Table 4-2. System Capacity Annual Cost per kW

	Fixed charge	O&M	$/kW	Annual cost, $/kW
Distribution	0.15	0.02	231	39.27
Transmission	0.145	0.02	141	23.27
Generation	0.143	0.02	611	99.59
Levelized annual cost of system capacity				$162.13

where a = inflation rate
    i = minimum acceptable return
    n = transformer life cycle in years
    CRF_n = uniform series capital recovery factor at interest i for n years.

For example, a typical escalation factor at 9 percent minimum acceptable return with a 4 percent levelized inflation rate over 30 years would be 1.47 or 1.84 for a 6 percent rate for a 30-year period. In order to account for inflation, an escalation factor must be applied to the energy cost components. If the inflation rate were uniform at 4 percent and the first-year energy cost was $0.0175/kWh, the levelized annual cost of energy would be calculated as follows:

$$\text{Levelized annual energy cost} = (\$0.0175 \times 1.47) = \$0.026$$

As often is the case, the inflation rate is not uniform, and the escalation of the energy cost must be calculated on a year-by-year basis over the life of the transformer. Table 4-3 illustrates how to determine the energy cost component of a transformer over the 30-year life of the transformer.

Levelized Annual Fixed Charge Rate (FC). First, determine the levelized annual fixed charge rate (see App. X) by adding the following investment-related components: minimum acceptable return (a composite value comprised of equity return and debt return), the book depreciation (including retirement dispersion, where applicable), income taxes (adjusted for accelerated depreciation and investment tax credit), and local property taxes and insurance. This term converts the levelized annual cost of losses into a capitalized value. It is often referred to as the annual cost ratio. It can be multiplied by the bid price to convert the cost of the transformer into an annual cost, or it can be divided into the annual cost of transformer losses to convert the annual cost of transformer losses into a capitalized cost. The total owning cost method discussed earlier in this chapter requires converting all costs to capitalized values. The following example from 1981 illustrates how to determine the fixed annual charge rate:

Minimum acceptable return	= 0.100
Book depreciation	= 0.250
Income taxes	= 0.150
Local property taxes and insurance	= 0.020
Fixed charge rate	= 0.160

Table 4-3. Levelized Energy Cost

Year	Energy cost, $/kWh (4% inflation year 5)	Single-payment present-worth factor @ 9%	Present worth energy cost
1	0.0175	0.91730	0.0161
2	0.0180	0.84168	0.0152
3	0.0190	0.77218	0.0147
4	0.0205	0.70843	0.0145
5	0.0220	0.64993	0.0143
6	0.0235	0.59627	0.0140
7	0.0250	0.54703	0.0137
8	0.0265	0.50187	0.0133
9	0.0280	0.46043	0.0129
10	0.0293	0.42241	0.0124
11	0.0308	0.38753	0.0119
12	0.0322	0.35553	0.0114
13	0.0338	0.32618	0.0110
14	0.0352	0.29925	0.0105
15	0.0368	0.27454	0.0110
16	0.0380	0.25187	0.0096
17	0.0395	0.23107	0.0091
18	0.0410	0.21199	0.0087
19	0.0425	0.19449	0.0083
20	0.0440	0.17843	0.0079
21	0.0453	0.16370	0.0074
22	0.0465	0.15081	0.0070
23	0.0483	0.13778	0.0067
24	0.0499	0.12640	0.0063
25	0.0512	0.11597	0.0059
26	0.0528	0.10639	0.0059
27	0.0542	0.09761	0.0056
28	0.0556	0.08955	0.0053
29	0.0572	0.08216	0.0047
30	0.0585	0.07536	0.0044
		Total present worth (TPW) =	0.2979

NOTE: Levelized energy cost = $TPW \times$ 30-year $CRF = 0.2979 \times .0973 = \$0.029$

Taking the fixed charge rate of 0.16, the system capacity cost of $162.13 from Table 4-2 and the energy cost of $0.029 from Table 4-3 derived in the 1981 EEI report, the cost of no-load losses, or factor A, can be calculated as follows:

$$A = \frac{SC + (EC \times 8760)}{FC} \times \frac{1}{1000} = \text{cost of no-load loss in } \$/W$$

where SC = \$162.13
EC = \$0.029
8760 = hours per year
FC = 0.16
1/1000 = conversion from kilowatts to watts

$$A = \frac{\$162.13 + (\$0.026 \times 8760)}{0.16} \times \frac{1}{1000} = \$2.77/\text{watt}$$

This compares to a more recent no-load loss value for distribution transformers. The costs of no-load losses vary from utility to utility. In a 1994 survey of 90 utilities, the Oak Ridge National Laboratories found that the average no-load loss value for distribution transformers was \$3.43 per watt. The following is an example of the value of no-load losses in distribution transformers from E-2 of the *Feasibility of Replacing or Upgrading Utility Distribution Transformers During Routine Maintenance* (1995) by Oak Ridge National Laboratories.

Example A utility has a peak load of 10,000 MW that it serves only 1 hour during the year and a base load of 4000 MW that it must meet for all 8760 hours of the year. Generally, the 4000-MW base load will be served by capacity generation sources, such as coal or nuclear power. The cost of the system base load, including interest during construction, is \$1792. This is the overnight construction for a bituminous, medium-sulfur coal-fired power plant, plus allowance for funds used during construction. However, this has been adjusted down to \$1277 because it has been assumed that on a national basis, base load capacity is not in short supply for 5 years. The adjustment is based on the \$1792 cost 5 years in the future discounted to a present value at 7 percent. A further downward adjustment was made to \$958 by multiplying by 0.75 to reflect the adjustment in incremental capacity that would occur in changing the size of a new plant. For instance, the average cost of an additional 50 MW added to a 600-MW plant would be less than the average cost per megawatt of the full plant. The capital carrying charge is based on a real interest rate of 7 percent and a 30-year depreciation life plus 1.5 percent for insurance and retirement dispersion, for a total of 9.6 percent. The levelized incremental production cost used to determine the A factor is 1.9 cents/kWh for starting fuel cost plus 1.7 percent annual real rate of escalation plus 0.6 cents/kWh for incremental operation and maintenance cost, for a total of 2.8 cents/kWh.

solution

$$A = \frac{SC + (EC \times 8760)}{FC} \times \frac{1}{1000} = \text{cost of no-load loss in \$/W}$$

where SC = \$958 × .0956 = \$91.58/year
EC = \$0.028
FC = 0.0956

$$A = \frac{\$91.58 + (\$0.28 \times 8760)}{0.0956} \times \frac{1}{1000} = \$3.53$$

After determining the value for no-load losses, or A factor, it is necessary to determine the corresponding value for load losses, or B factor.

Methods for Valuing Distribution Transformer Load Losses (or B Factor)

Calculate the value for distribution load losses, or B factor, according to the following formula:

$$B = \frac{SC \times RF + 8760 \times EC \times LF}{FC} \times \frac{(PL)^2 \times 1 = \$120}{1000} \qquad (4.3)$$

where RE = peak loss responsibility factor. RF is the composite responsibility factor that reduces the system capacity requirement for load losses, since the peak transformer losses do not necessarily occur at peak time.

LF = annual loss factor. LF is the ratio of the annual average load loss to the peak value of the load loss in the transformer.

PL = uniform equivalent annual peak load. PL is the uniform series of annual peak transformer loads equivalent to the actual nonuniform series of annual transformer peak loads over the transformer life cycle that produces the same total present worth of losses expressed in per unit of nameplate rating.

Transformer life cycle is defined as the useful life of the asset and is usually assumed to be 30 to 35 years. The process for solving this equation is illustrated in Fig. 4-3.

Common to both the value of no-load and load losses is the system capacity cost SC, while calculation of the value of load loss requires knowing the peak responsibility factor RF.

Peak Responsibility Factor RF. The peak responsibility factor was introduced briefly in Chap. 3. It accounts for the fact that the peak loading on a distribution transformer usually does not occur at the same time as the peak loading on the various components of the utility's system (i.e., generation, transmission, and distribution). The responsibility factor RF is the ratio of the system components' peak load over the correspond-

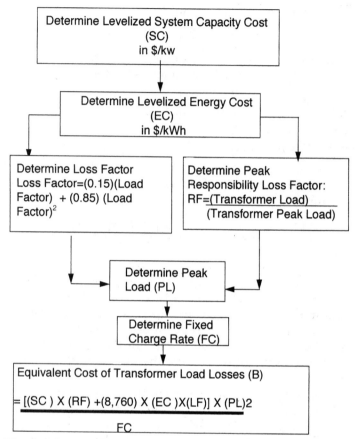

Figure 4-3. Determining cost of transformer load losses.

ing transformer's peak load. A good approximation of the composite peak loss responsibility factor is a simple average.

Annual Loss Factor LF.   The annual loss factor is used to determine the annual losses in the distribution transformers. It is usually derived from the load factor. Recalling from Chap. 3 that the load factor is the ratio of the average load to the peak load, the loss factor is determined from the following formula:

$$\text{Loss factor} = 0.15 \,(\text{load factor}) + 0.85 \,(\text{load factor})^2 \qquad (3.8)$$

The load factor is not a difficult term to determine in a distribution system. Metering transformer monitoring stations can be used to determine the load factor using the following formula:

$$\text{Annual load factor} = \frac{\text{annual kWh consumption}}{\text{annual peak kW}} \qquad (4.4)$$

If these data are not available from the field, they can be obtained by metering a sampling of distribution transformers.

Uniform Annual Peak Load (PL). Uniform annual peak load is the levelized peak load per year over the life of the transformer. This requires a three-step process: (1) determine the peak load for 1 year, (2) project peak load into the future, and (3) take the present worth of each peak load and levelize it over the life of the transformer. An example developed in 1981 by the utility industry report illustrates how to determine the uniform annual peak load *PL*.

$$PL = \left(\left\{ \sum_{j=1}^{n} \left[b(1 + g)^{(j-1)} \right]^2 SPPWF_i^n \right\} CRF_i^n \right)^{1/2} \qquad (4.5)$$

where n = number of years to reach changeout load
$\quad i$ = minimum acceptable return
$\quad b$ = the initial transformer load in per unit of nameplate rating (end of year 1)
$\quad g$ = the annual peak load growth experienced by the distribution transformer per unit
$\quad n$ = the number of years required for the transformer to attain its changeout load level
$SPPWF_i^n$ = single-payment present-worth factor = $\left[\dfrac{1}{(1 + i)^n} \right]$

CRF_i^n = capital recovery factor = $\left[\dfrac{i}{(1 + i)^n - 1} + i \right]$

In this example, distribution transformers are installed at 80 percent of nameplate rating and are changed out at 150 percent of nameplate rating with a 30-year life. One changeout is required during the life cycle, and the annual load growth percentage g is 4.6 percent. The per unit uniform annual peak load can be determined by first calculating the annual load growth percentage using the exponential growth equation. Table 4-4 shows the year-by-year solution for levelized peak load. With the total present worth of 11.717, the uniform equivalent annual peak load *PL* can be calculated:

$$PL = [(\text{total present worth})(CRF_{9\%})]^{1/2} = \sqrt{11.717(.09734)} = 1.068$$

Based on these calculations, the value of load losses, or *B* factor, can be calculated as follows:

$$B = \frac{SC \times RF + 8760 \times EC \times LF}{FC} \times \frac{(PL)^2 \times 1}{1000}$$

Table 4-4. Uniform Equivalent Annual Peak
Load Factor

Year j	Per unit nameplate load squared in year j	SPPWF $(i = 9\%)$	Present worth of PL^2
1	0.64	0.91743	0.5872
2	0.7	0.84168	0.5892
3	0.766	0.77218	0.5915
4	0.838	0.70843	0.5937
5	0.917	0.64993	0.5960
6	1.003	0.59627	0.5981
7	1.098	0.54703	0.6006
8	1.201	0.50187	0.6027
9	1.314	0.46043	0.605
10	1.4838	0.42241	0.6074
11	1.558	0.38753	0.6038
12	1.721	0.35553	0.6119
13	1.885	0.32618	0.6148
14	2.060	0.29925	0.615
15	2.250	0.27454	0.6177
16	0.640	0.2518	0.1612
17	0.700	0.23107	0.1617
18	0.7660	0.21199	0.1624
19	0.838	0.19449	0.1630
20	0.197	0.17843	0.1636
21	1.003	0.16370	0.1642
22	1.098	0.15081	0.1649
23	1.201	0.13778	0.1656
24	1.314	0.12640	0.1661
25	1.438	0.11597	0.1668
26	1.558	0.10639	0.1658
27	1.721	0.9761	0.1680
28	1.885	0.08955	0.1688
29	2.060	0.08216	0.1692
30	2.250	0.07536	0.1696
		Total present worth =	11.7170

SOURCE: Courtesy of Edison Electric Institute.

where $RF = 0.53$
$\quad\quad\; LF = 0.12$
$\quad\quad\; PL = 1.068$

Recalling from the value of no-load loss calculations that $SC = 162.13$ and $EC = 0.029$, then

$$B = \frac{162.13 \times 0.53 + 8760 \times 0.029 \times 0.12}{0.16} \times (1.068)^2 \times \frac{1}{1000}$$

$$= \frac{(0.1164)(1.14)(1)}{0.16} = \frac{0.1327}{0.16} = \$0.83$$

This compares to a more recent load loss value for distribution transformers. The costs of no-load losses vary from utility to utility. In a 1994 survey of 90 utilities, the Oak Ridge National Laboratories found that the average load loss value for distribution transformers was $1.09, with a standard deviation of $.90 per watt.

The preceding discussion provides a straightforward procedure for calculating the cost of no-load and load losses in distribution transformers from a utility's perspective. Before discussing the cost of no-load and load losses in distribution transformers from an industrial and commercial perspective, it makes sense to discuss the calculation of the cost of no-load and load losses in substation transformers.

Substation Transformers

Calculating the value of no-load and load losses for substation transformers is somewhat similar to calculating the value of no-load and load losses for distribution transformers. The three major differences are (1) the cost of the distribution system is not included in the capacity cost; (2) the cost of load losses must be calculated for the various stages of cooling and loading in a substation transformer; and (3) the cost of auxiliary losses must be included in the cost of losses for substation transformers.

The utility standard for evaluating losses on substation transformers is the *IEEE Loss Evaluation Guide for Power Transformers and Reactors* (IEEE Standard C57.120-1991). The formulas contained in this standard are similar to those for distribution transformers except that the symbols have been changed. In order to avoid confusion, this book will use the same symbols for calculating the cost of losses in substation transformers as were used in calculating the cost of losses in distribution transformers. The only exception in symbols will occur where a calculation is specific to substation transformers, as in auxiliary losses. The various terms and corresponding symbols for the cost of distribution transformer losses are assumed to be the same for substation transformers.

Methods for Valuing Substation Transformer No-Load Losses (A Factor). In the case of substation transformers, determining the value of no-load losses is the same for distribution transformers except for the need to determine the availability factor AF. The availability factor affects the cost of energy consumed in the substation transformer. The IEEE Loss Evaluation Guide for Power Transformers and Reactors (IEEE Standard C57.120-1991) recommends the use of an availability factor to account for the substation transformer being down throughout the year for routine maintenance. Consequently, in the formula for the cost of no-load losses, the number of hours in a year (8760) is multiplied by the availability factor to account for the substation transformer not being available every hour throughout the year. The formula for the cost of no-load losses in substation transformers thus becomes

$$A = \frac{SC + (EC \times AF \times 8760)}{FC} = \text{cost of no-load loss in \$/kW}$$

Notice that the cost of no-load losses is in dollars per kilowatt for substation transformers rather than dollars per watt for distribution transformers. This is so because the losses in a substation transformer are quite a bit larger than the losses in a distribution transformer. It is no longer necessary to divide the cost of losses by 1000 in the case of substation transformers.

The example in the IEEE standard illustrates how to calculate the cost of no-load losses:

Example Assuming a fixed charge rate of 0.193, a system capacity cost of $172, and an energy cost of $0.059/kWh, the cost of no-load losses, or A factor, can be calculated as

$$A = \frac{SC + (EC \times 8760)}{FC} = \text{cost of no-load loss in \$/kW}$$

where SC = $172/kW-yr
 EC = $0.059/kWh
 8760 = hours per year
 FC = 0.193
 AF = 0.97

$$A = \frac{\$172 + (\$0.059 \times 0.97 \times 8760)}{0.193} = \$3497 \ \$/kW$$

Methods for Valuing Substation Transformer Load Losses (B Factor). The formula for the value of load losses in substation transformers contained in the IEEE Loss Evaluation Guide for Power Transformers and Reactors (IEEE Standard C57.120-1991) is quite similar to the corresponding formula for the value of load losses in distribution transformers. The only difference is the method for calculating the contribution of energy cost to the total value of load losses. In the case of the distribution transformer method, the energy cost contribution is calculated by multiplying energy cost *EC* by the loss factor *LF* times the uniform annual peak load *PL* squared, while IEEE Standard C57.120-1991 for substation transformers calculates the energy cost contribution by multiplying the energy cost *EC* by the transformer loading factor *TLF* squared. *Transformer loading factor* is defined in IEEE Standard C57.120-1991 as "the root-mean-square value of the predicted loads of the power transformer over a representative yearly period is an equivalent load. This equivalent load, in MVA, divided by the rating at which the load losses are guaranteed and tested." For the purposes of this book, and to avoid confusion, again, the same formulas used in the 1981 EEI report for the value of distribution transformer losses will be used for the value of substation transformer losses.

$$B = \frac{SC \times RF + 8760 \times EC \times LF}{FC} \times (PL)^2$$

$$= \text{load loss cost } \$/\text{kW}$$

Continuing the example for the value of substation transformer losses contained in IEEE Standard C57.120-1991, we have the following.

Example Assuming a fixed charge rate of 0.193, a system capacity cost of \$172, and an energy cost of \$0.059, peak per unit load *PL* of 1.67, and a loss factor of 0.09,

$$B = \frac{SC \times RF + 8760 \times EC \times LF}{FC} \times (PL)^2$$

where $SC = \$172/\text{kW-yr}$
$EC = \$0.059/\text{kWh}$
$RF = 0.92$
$PL = 1.67$
$LF = 0.09$
$FC = 0.193$

$$B = \frac{172 \times 0.92 + 8760 \times 0.059 \times 0.09}{0.193} \times 1.67^2$$

$$= \$2958/\text{kW}$$

Substation transformers often contain fans and pumps that allow them to be loaded at additional ratings. Usually the loading is in two stages. These stages of cooling have auxiliary losses and a corresponding value of auxiliary losses associated with operation of the fans and pumps.

Methods for Valuing Substation Transformer Auxiliary Losses.
The value of auxiliary losses is the same as the calculation of the value of
no-load losses, except that cooling equipment does not run for the full
year. The value of auxiliary losses D can then be determined by taking the
energy component for the no-load losses and multiplying by an availability factor equal to the ratio of the auxiliary equipment operating time for
the year in hours divided by the total number of hours in a year the transformer is used. The following formula illustrates this calculation:

$$D = \frac{SC + (EC \times \text{running time/transformer use})}{FC} \quad (4.6)$$

Using the example in the IEEE Standard C57.120-1991, the value of
auxiliary losses for stage 1 cooling equipment power, assuming that
stage 1 operates 3000 hours a year and the transformer is used 8497
hours in a year, is

$$\text{Value of stage 1 auxiliary losses} = \frac{172 + (500 \times 3000/8497)}{0.193}$$

$$= \$1804/\text{kW}$$

Assuming that the stage 2 cooling equipment operates 1000 hours per
year, the value of stage 2 auxiliary losses can be calculated as

$$\text{Value of stage 2 auxiliary losses} = \frac{172 + (500 \times 1000/8497)}{0.193}$$

$$= \$1196/\text{kW}$$

Commercial and Industrial Perspective

According to the 1996 Oak Ridge National Laboratory report entitled,
*Determination Analysis of Energy Conservation Standards for Distribution
Transformers*, "Utilities purchase and operate about 60 percent of all
transformer capacity. The remaining 40 percent of transformer capacity
is owned by commercial and industrial establishments." Most commercial and industrial users of transformers do not evaluate losses,
even though they purchase a significant amount of the transformer
capacity throughout the United States. Why don't commercial and
industrial users evaluate transformer losses? Most of them purchase
transformers on a small scale and do not develop their transformer
expertise. In fact, most of them do not purchase the transformers
directly from the manufacturer but buy them through a manufactur-

er's representative. This is why a commercial and industrial purchaser of transformers could benefit from this book.

Rather than go through the extensive research and analysis that a utility does in developing its transformer cost of losses, commercial and industrial (C&I) transformer purchasers can use a more simplified but effective approach for evaluating transformer losses. This involves using the same formulas for the cost of no-load and load losses but using the demand and energy rate the utility charges them for the capacity cost and energy cost. The Oak Ridge National Laboratory report entitled, *Determination Analysis of Energy Conservation Standards for Distribution Transformers*, says: "An analysis from the C&I user perspective indicated that the evaluation losses for C&I firms would tend to be somewhat lower than those assumed in this analysis" (i.e., utilities). The difficulty is in predicting how much the utility's rates will change in the future. It therefore becomes paramount for the transformer purchaser to understand how to perform an economic sensitivity analysis as well as project future industry energy trends.

Industry Energy Trends

According to the U.S. Department of Energy *1996 Annual Energy Outlook* [DOE/EIA-0202 (96/2Q)], world energy consumption in 1990 totaled 345 quads of Btu. Between 1970 and 1990, consumption increased by almost 140 quads of Btu, reflecting an annual growth rate of 2.6 percent. Growth in energy consumption is expected to slow between 1990 and 2010, averaging 1.6 percent, compared with 2.6 percent in the previous two decades.

Electricity shows the most rapid growth of all worldwide energy components. In industrialized countries like the United States, electricity consumption grew at 3.7 percent per year between 1970 and 1990, while overall energy consumption grew by 1.6 percent per year. In the United States, electricity demand in the commercial sector is projected to rise by 1.7 percent in 1996 and by 1.3 percent in 1997, due primarily to expanding employment. The industrial demand for electricity is projected to grow more slowly in 1996 at 0.7 percent due to slower economic growth than in 1995 and to grow by 1.9 percent in 1997, reflecting a resurgence in economic growth. Figure 4-4 shows the increase in U.S. electricity demand from 1994 to 1997.

Between 1990 and 2010, world electrical consumption growth is expected to slow to 1.7 percent per year. Electricity consumption in nonindustrialized countries grew by approximately 5 percent per year between 1970 and 1990 and is expected to grow at a greater rate between 1990 and 2010. Electricity consumption has grown much

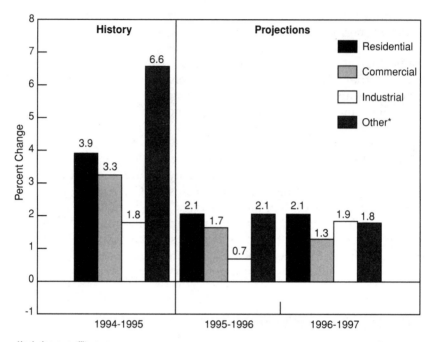

*Includes nonutility own use
Mid World Oil Price Case
Sources: Second Quarter 1996 STIFS database.

Figure 4-4. U.S. electricity demand. (*Courtesy of the United States Department of Energy 1996 Annual Energy Outlook, DOE/EIA-0202, 96/20.*)

faster in nonindustrialized Asia, where the 1990 level is 5 times the 1970 level. In Asia, electricity consumption is projected to double from 1990 to 2010. However, how does this growth in electrical energy consumption affect the present and future cost of electric power?

Electric Power Cost Trends

Electric power cost was relatively constant during the decade of the 1960s and the beginning of the 1970s. During the energy crisis of the early 1970s, the cost of electric power to industrial and commercial users increased by 11 to 12 percent per year. However, as with any projection, there is an element of uncertainty. These uncertainties include

1. World supply and demand
2. Import price and availability

3. Size of resources
4. Economic health
5. Government policy
6. Deregulation
7. Increased competition
8. Increased use of electricity

How does one factor these cost trends and uncertainties into the value of transformer load and no-load losses? One method to overcome uncertainty is to perform a sensitivity analysis. *Sensitivity analysis* involves analyzing the various assumptions in developing the cost of transformers to determine if any change would affect the decision as to what transformer to purchase. This approach will be discussed in more detail in Chap. 6, "Transformer Economics."

5
Transformer Cost

Transformer cost is a major factor determining what transformer to buy. Along with evaluating the cost of losses, the transformer purchaser needs to evaluate transformer cost. There is a relationship between the transformer cost and the amount of transformer losses. In fact, just as a transformer purchaser seeks to optimize losses on its power system by purchasing cost-effective and energy-efficient transformers, transformer manufacturers seek to optimize the losses in the transformer design. The components in a transformer design, construction, and installation all affect its cost. For example, amorphous metal core transformers have 75 percent less no-load loss than a silicon steel core transformer but cost 25 percent more than a silicon steel core transformer. This will be discussed in more detail in Chap. 9, "Amorphous Steel Core Distribution Transformers."

How do transformer losses affect the cost to design, construct, and install transformers? The transformer manufacturer incorporates the cost of losses into the cost of the transformer in optimizing the transformer design. The transformer designer has various formulas applied to a computer program to determine transformer cost. These formulas include the material, labor, and overhead that go into making a transformer. The transformer designer must weigh the cost of these factors as he or she tries to meet the customer's specifications.

How to determine the transformer cost as it relates to transformer losses is not an easy task. In the evaluation of transformer design, after the A and B factors the next major component in the total owning cost formula is the cost of the transformer, or C factor, similar to the formula in Chap. 1:

$$TOC = NLL \times A + LL \times B + C + D \qquad (5.1)$$

where *TOC* = capitalized total owning cost
    NLL = no-load loss in watts
       A = capitalized cost per rated watt of *NLL* (*A* factor)
     LL = load loss in watts at the transformer's rated load
       B = capitalized cost per rated watt of *LL* (*B* factor)
       C = the initial cost of the transformer, including transportation, sales tax, and other costs to prepare it for service
      D = cost of auxiliary cooling losses in the case of transformers with auxiliary cooling

Customer Perspective

The transformer purchaser provides the values of the *A* and *B* factors in the total owning cost formula to the transformer manufacturer. The transformer manufacturer integrates the values of the *A* and *B* factors into other costs of the transformer design. The costs of the *A* and *B* factors become part of the design specification. This includes the specified transformer overload capability, insulation thermal capability, short-circuit capability, transformer impedance, specified phasing, in-service conditions, voltage regulation, and the basic impulse level. The transformer purchaser sometimes specifies the transformer efficiency but usually leaves it up to the transformer manufacturer based on the values of the *A* and *B* factors. Tables 5-1 through 5-3 illustrate how the value of the *A* and *B* factors affect the cost of the transformer.

Design 3, the most expensive distribution transformer design, is the most cost-effective of the four designs when the capitalized cost of no- .

Table 5-1. Economic Tradeoffs for Hypothetical Transformer Designs (Assuming Cost per Watt of No-Load Loss is $4 and Cost per Watt of Load Loss Is $1)

Transformer values	Design 1	Design 2	Design 3	Design 4
No-load losses (W)	100	80	50	120
Load loss (W)	220	280	290	310
Initial cost	$490	$475	$580	$350
Cost of no-load losses	$400	$320	$200	$480
Cost of load losses	$220	$280	$290	$310
Total owning cost	$1110	$1075	$1070	$1140
Rank	3	2	1	4

SOURCE: Courtesy of ORNL.

Table 5-2. Economic Tradeoffs for Hypothetical Transformer Designs (Assuming Cost per Watt of No-Load Loss is $2.50 and Cost per Watt of Load Loss Is $0.75)

Transformer values	Design 1	Design 2	Design 3	Design 4
No-load losses (W)	100	80	50	120
Load loss (W)	220	280	290	310
Initial cost	$490	$475	$580	$350
Cost of no-load losses	$250	$200	$125	$300
Cost of load losses	$165	$210	$218	$233
Total owning cost	$905	$885	$923	$883
Rank	3	2	4	1

SOURCE: Courtesy of ORNL.

Table 5-3. Economic Tradeoffs for Hypothetical Transformer Designs (Assuming Cost per Watt of No-Load Loss is $2.80 and Cost per Watt of Load Loss Is $0.75)

Transformer values	Design 1	Design 2	Design 3	Design 4
No-load losses (W)	100	80	50	120
Load loss (W)	220	280	290	310
Initial cost	$490	$475	$580	$350
Cost of no-load losses	$280	$224	$140	$336
Cost of load losses	$165	$210	$218	$233
Total owning cost	$935	$909	$938	$919
Rank	3	1	4	2

SOURCE: Courtesy of ORNL.

load losses is $4 per watt and the capitalized cost of load losses is $1 per watt for a 30-year study period assuming a 9.56 percent carrying charge rate, while design 4, the least expensive design, is the most cost-effective of the four designs when the cost of no-load losses drops to $2.80 per watt and the cost of load losses drops to $0.75 per watt. This clearly shows that increased efficiency and increased initial cost can only be offset by an increase in the cost of no-load and load losses. Even though the effect of the transformer losses on the cost of the transformer is similar for substation and distribution transformers, distribution transformers and substation transformers will be dealt with separately.

Distribution Transformers

The effect of no-load and load losses on the cost of distribution transformers is best illustrated in a three-dimensional graph developed by Oak Ridge National Laboratories using data provided by the National Electric Manufacturers Association. Figure 5-1 illustrates how the transformer designer's efforts to obtain certain values of no-load and load losses affect the volume, weight, and amount of materials in a transformer and consequently the transformer cost for a given kilovoltampere capacity and voltage.

As can be seen from this graph, the cost of the distribution transformer is sensitive to the magnitude of the no-load and load losses. The distribution transformer designer must juggle several factors in order to obtain an optimal design based on the customer's specification. The no-load losses are interrelated. The amount of core that determines the no-load loss affects the amount of copper that affects

ORNL-DWG 95M-10063

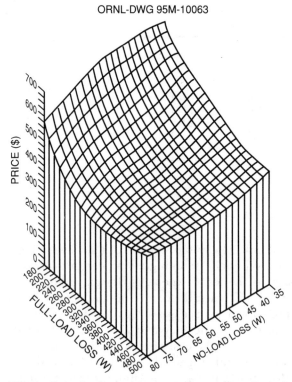

Figure 5-1. Surface cost versus losses for a typical 25-kVA distribution transformer. (*ORNL prepared using data supplied by the National Electric Manufacturers Association.*)

the load losses, and vice versa. Still, it is helpful to deal separately with the effect of no-load and load losses on transformer cost.

No-Load Loss Effect on Transformer Cost. Transformer no-load loss tests show that by increasing the number of turns in the transformer coil, the no-load losses decrease proportionately. This is so because increasing the number of turns in either the primary or secondary winding increases the magnetizing force. The increased magnetizing force increases the flux density and results in less no-load loss. However, increasing the number of turns in the windings increases the length of the winding conductor and thus the resistance of the winding. This increases the winding or load loss. Therefore, the transformer designer must balance the no-load and load losses in that one affects the other.

Besides, the distribution transformer designer also must take into account the telephone influence factor (TIF). The TIF is a way of reducing the effects of harmonic voltages generated in the transformer coil affecting telephone communication. Older power lines and telephone lines often were strung on the same pole and were in close proximity to each other. The old Rural Electrification Administration set limits on harmonic voltages in distribution transformer cores several years ago. Many utilities require the manufacturer to limit the harmonic voltages. The transformer manufacturer has to design the core so as not to exceed the TIF requirements of the customer.

Another factor that affects the size of the core is the need to reduce 120-Hz transformer "noise." This noise is caused by contraction and expansion of the core while being subjected to a magnetic field. Just as increasing the core cross-sectional area results in reducing core or no-load losses, it also reduces transformer noise. The increased core cross-sectional area increases the cost of the transformer, however.

Load Loss Effect on Cost. It would seem that increasing the size of conductors would provide lower losses. Just as lower no-load losses increase load losses, lower load losses increase no-load losses. This is so because transformers with larger coils require larger cores with corresponding increases in no-load losses. This is why the rule of thumb of having no-load losses equal to the load losses at the transformer's rated loading is still a good one.

It is helpful to understand the relationship between losses and transformer costs from the manufacturer's perspectives. Then the user of the transformer can appreciate the difficulty and challenges that the transformer manufacturer faces in trying to achieve certain loss values and still be cost-competitive. Edward Boyd of Westinghouse wrote an article in the May 1980 issue of *Electric Light and Power* entitled, "Modified Standard Transformers Best Hedge Against Losses?" In this article he presented the following simple formulas of the relationship between distribution transformer costs and losses:

$$TW/\# = f(B) = f\left(\frac{k_1}{A} \times \frac{E}{n}\right) \qquad (5.2)$$

$$\text{Core weight} = g(A, \text{conductor weight}) \qquad (5.3)$$

$$\text{No-load loss} = k_2 \times \text{core weight} \times TW/\# \qquad (5.4)$$

$$\text{Load loss} = h(1/\text{conductor weight}, A, n/E) \qquad (5.5)$$

where A = core cross-sectional steel area
$\quad B$ = maximum flux density in core
$\quad E/n$ = volts per turn = rms voltage of any winding in the coil divided by the number of turns in that winding
$\quad f, g, h$ = nonlinear functions
$\quad k_1$ = constant varying with unit of measure chosen for A
$\quad k_2$ = constant accounting for differences in $TW/\#$ from Eq. 5.2 and actual $TW/\#$ in a wound core
$\quad TW/\#$ = watts of no-load loss per pound of core weight

As can be seen from these formulas, losses have an effect on the physical design of the transformer. This is why it is important for the transformer purchaser to communicate to the manufacturer its loss evaluation criteria. This allows the manufacturer to design a transformer that minimizes the transformer cost and losses, a difficult balancing act that is best accomplished by clear communication between the manufacturer and user of transformers. Clearly, the transformer economics used by the transformer purchaser are critical to the whole equation.

6
Transformer
Economics

An understanding of transformer economics is necessary to weigh the transformer cost against the benefits of transformer efficiency. As with all economic analyses, the time value of money over the life cycle of the alternatives needs to be evaluated. Efficiency improvement and loss savings occur over time and must somehow be compared with the initial cost of purchasing and installing the transformer. There are basically three standard methods for evaluating alternative transformer choices:

1. Equivalent investment cost
2. Levelized annual cost
3. Present-worth method

Each one of these methods will be discussed as it applies to the initial cost of the transformer and the costs of no-load and load losses. Each method must be applied to the total owning cost formula in such a way that the various parts of the formula are compared on an equitable basis. How to determine the transformer economic method as it relates to the total owning cost formula is a matter of company policy and personal preference. The economic method should have no effect on the decision as to what transformer to buy. In evaluating transformer design, the transformer economic method is applied to the components of the total owning cost, Eq. 5.1:

$$TOC = NLL \times A + LL \times BC + C + D \qquad (5.1)$$

The resulting numbers derived from any one of these methods applied to the total owning cost formula can be used to choose the most cost-effective and energy-efficient transformer. Some transformer purchasers, because of the uncertainty of the assumptions and values used in the total owning cost formula, choose to use the band of equivalent (*BOE*) method. The band of equivalent method involves picking a transformer that fits a band of equivalent value calculated from the total owning cost formula. Another way transformer purchasers deal with the uncertainty of the values in the total owning cost formula is to perform a sensitivity analysis. A sensitivity analysis involves determining how sensitive the total owning cost is to assumptions and values in the total owning cost formula. How to use the band of equivalent method and perform a sensitivity analysis will be discussed later in this chapter.

Customer Perspective

The purchaser of transformers wants to make a decision as to what transformer to purchase with a minimum amount of difficulty. An understanding of the various methods for performing an economic analysis of a transformer purchase is essential to reducing that difficulty. The transformer purchaser needs to decide on an economic methodology that he or she is most comfortable using. It is important that the transformer purchaser consistently use the economic method throughout the transformer purchasing decision-making process. This allows the transformer manufacturer to design a transformer that meets the needs and values of the customer. Figure 6-1 illustrates the process that the transformer purchaser needs to go through in choosing and using an economic method for evaluating a transformer purchase.

Life Cycle

In performing any kind of economic analysis, it is necessary to take into account the life-cycle cost of the transformer. Life-cycle costing is the fundamental concept used in deriving the total owning cost formula. It involves calculating the total cost of ownership over the life span of the transformer. Only then can the cost of losses be compared with the costs of purchasing, operating, and maintaining the transformer. So what is the life span of a transformer? Is it based on its expected life before failure? Or is it based on its expected life before replacement?

Most utilities and some commercial and industrial transformer owners use the expected life of the transformer to evaluate the losses.

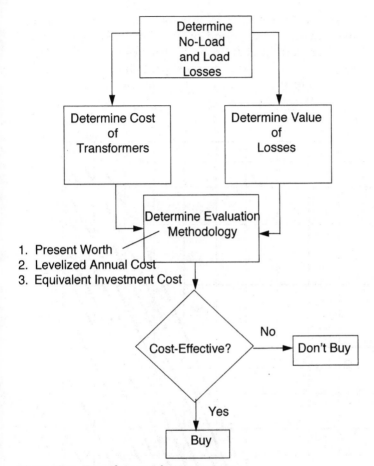

Figure 6-1. Transformer selection process.

However, if the transformer is replaced before it reaches its full life due to failure, overloading, or high losses, the expected replacement time should be incorporated into the comparison.

There are many factors that affect a transformer's life. Anything that affects the insulating strength inside the transformer reduces transformer life. Such things as overloading the transformer, moisture in the transformer, poor-quality oil or insulating paper, and extreme temperatures affect the insulating properties of the transformer. Increased temperature is the major cause of reducing transformer life. Most transformers are designed to operate for 20 years at the nameplate rating. While transformers that are loaded below the nameplate rating have lives longer than 20 years, transformers loaded above the name-plate rating over an extended period of time have lives less than 20 years. Transformer manufacturers have developed graphs of the effects of temperature and loading on transformer life. Figure 6-2 illustrates how temperature and loading affect transformer life.

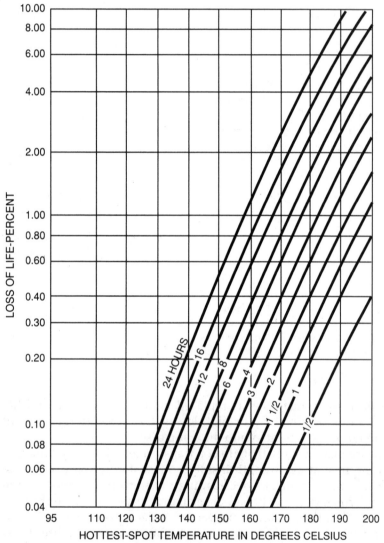

Figure 6-2. Hottest spot temperature versus loss of transformer life. (*Courtesy of IEEE C57.91-1995.*)

Transformers may be replaced before their 20-year or more life span. They may be replaced due to failure from overloading or damage caused by nature or system operating conditions. Or they may be replaced for energy-efficiency improvement reasons. As new and more energy-efficient transformers are developed and the cost of energy increases, it may be cost-effective to replace a transformer before it reaches its life span. This determination is usually based on comparing the energy savings resulting from early replacement with the cost of a new transformer. Chapter 7, "Transformer Replacement," discusses in more detail how to decide if it is cost-effective to replace a transformer with a more energy-efficient one.

Whether a transformer reaches the end of its life due to failure or replacement, the transformer life varies from one user to another. Based on the Oak Ridge National Laboratory report, the national average for utility distribution transformers is 31.95 years. Therefore, a life of 30 years is a good assumption.

Time Value of Money

Each one of the economic analyses involves the time value of money (see App. F). What is the time value of money? The *time value of money* means that money increases in value over time depending on the return on the investment. If money is deposited to an interest-bearing savings account or money market, it could appreciate in value to the tune of about 5 percent per year, while if money is invested in a stock or mutual fund, it could appreciate 10 to 20 percent per year. Each company or person has an expectation about the time value of money. To ignore its value over time is not practical. Both the cost of the transformer and the cost of losses have time value.

An understanding of simple interest rate is essential to an understanding of the time value of money. Simple interest rate, carrying charge rate, minimum acceptable rate of return, cash flow diagrams, and the various engineering economic factors necessary to perform a transformer economic analysis are discussed in detail in the appendices. This chapter compares the various methods for evaluating transformer losses and initial costs and explains how to use these methods starting with the equivalent first-cost method.

Equivalent First Cost

The equivalent first-cost method is probably the most popular method of economic analysis. This is so because it is the most straightforward of all the methods. This method involves taking the total owning cost formula and adding the various components without any additional modifications to those components:

TOC = price + cost of the no-load losses + cost of the load losses

The price is the bid price of the transformer supplied by the manufacturer. This price requires no modification. The only caveat is that it be based on the same A and B factors in all cases.

The costs of no-load and load losses are calculated as follows:

Cost of no-load losses = $A \times$ no-load losses $\times$ loss multiplier

Cost of load losses = $B \times$ load losses $\times$ loss multiplier

The A and B factors, as described in Chap. 4, have been determined on the basis of equivalent first cost and need no further manipulation. This has the benefit of allowing the manufacturer to design the transformer to meet the A and B factors specified by the customer. While the no-load losses are the measured no-load losses for each transformer design, the load losses are the measured load losses at the rated nameplate loading. The loss multiplier accounts for the additional losses imposed on the transmission and distribution system as a result of the transformer losses. The other methods for evaluating losses are a modification of the equivalent first-cost method, starting with the levelized annual cost method.

Levelized Annual Cost Method

The levelized annual cost method (see App. K) has already been discussed in detail in Chap. 4. The 1981 utility standard method utilized the levelized annual cost method. It involves converting each component of the total owning cost formula into an annual levelized cost over the life of the transformer. Each component of the total owning cost formula remains the same except that it is converted into a levelized annual cost by multiplying by a single factor. This means that the initial cost, the A factor, the no-load losses, the B factor, and the load losses each must be multiplied by this single conversion factor. The levelized annual cost method is accomplished by multiplying the initial cost and the cost of losses by the carrying charge rate or fixed charge rate (FCR). The total owning cost formula for levelized annual cost changes the total owning cost formula for equivalent initial cost to the following:

$TOC = FCR \times$ (price + cost of the no-load losses + cost load losses)

As can be seen from this formula, as long as all transformer design alternatives apply the same *FCR* amount, the relationship between the alternatives will not be affected by this method. The decision as to the most cost-effective transformer will be the same in the equivalent first-cost method as in the levelized annual cost method. The present-worth method also will render the same decision, as can be seen from the following discussion.

Present Worth of Annual Revenue Requirements

The present-worth method requires taking each component of the total owning cost formula and referring them all back to a common date. This provides a comparison of various transformer designs and their corresponding efficiency values. The life of the transformer and the carrying charge rate (fixed charge rate) remain constant in this method.

In the present worth of annual revenue requirements method, the levelized annual cost method formula is multiplied by the uniform series present-worth (*USPW*) factor. This converts the equal annual cost values into a present-worth value. Thus the levelized annual price, the levelized annual cost of no-load losses, and the levelized annual cost of load losses are each converted into present-worth values. This can best be seen by looking at the total owning cost formula using this method:

$$TOC = USPW \times FCR \times (\text{price} + \text{no-load loss cost} + \text{load loss cost})$$

Again, because all the components of the total owning cost formula are multiplied by the uniform series present-worth factor, the relative relationship between transformer design alternatives is not changed. The cost-effectiveness of the various designs will be the same in the present-worth method as in the equivalent initial cost method and the levelized annual cost method. The uniform annual present-worth factor is discussed in App. F. The *USPW* can be determined from a table, assuming a minimum acceptable rate of return and the life of the transformer, while the auxiliary losses must be examined in the case of a substation transformer.

Evaluation of Auxiliary Losses in Substation Transformers

The evaluation auxiliary losses can be treated in a manner similar to the way the costs of load losses are handled, the difference being the fact that the auxiliary losses are operated in stages. In any case, the preferred method of evaluation is the equivalent first-cost method. The equivalent cost of the auxiliary losses can be converted to the levelized annual values by multiplying the cost of auxiliary losses by the carrying charge rate or the fixed cost rate. If the present-worth analysis method is preferred, then the levelized annual cost of the auxiliary losses can be multiplied by the uniform series present-worth factor. In all cases, the auxiliary evaluated losses are added to the total owning cost formula as follows:

$$TOC = \text{price} + \text{no-load loss cost} + \text{load loss cost} + \text{auxiliary loss cost}$$

Uncertainty

Uncertainty as to the validity of the total owning cost values is always a concern of the transformer customer. No matter what method is used to calculate total owning cost, its value is uncertain. This uncertainty is increased due to the lack of stability of rates in the utility industry and the long period of time for the transformer life-cycle analysis. This increased uncertainty causes concern about the various assumptions required in calculating the total owning cost. The uncertainty is compounded by the need to evaluate losses over the 30-year life of the transformer. Reliance on the assumed future value of the cost of energy and capacity, escalation and discount rates, transformer loading, and load and responsibility factors can affect the value of the total owning cost and the consequent decision to buy the most cost-effective transformer. How does the transformer customer deal with these uncertainties? Rather than rely on the absolute total owning cost values, many utilities instead are using the band of equivalent (*BOE*) approach.

Band of Equivalent (*BOE*)

In the band of equivalent approach, the purchaser picks a group of transformers that fall within a total owning cost band. From within that band, the lowest-cost transformer is chosen. This is based on the conclusion that by taking a band of equivalent approach the purchaser is dealing with the uncertainty in the value of the total owning cost. This approach often results in an actual reduction in the loss evaluation factors. This can best be illustrated by comparing design 1 to design 2 in Table 5-1 (which is presented again as Table 6-1). As can be seen from this table, changing from the total owning cost value to the

Table 6-1. Rank based on TOC versus BOE
(Assuming Cost per Watt of No-Load Loss Is $4
and Cost per Watt of Load Loss is $1)

Transformer values	Design 1	Design 2
No-load losses (W)	100	80
Load loss (W)	220	280
Initial cost	$490	$475
Cost of no-load losses	$400	$320
Cost of load losses	$220	$280
Total owning cost	$1110	$1075
Rank based on TOC	1	2
Rank based on BOE	2	1

SOURCE: Courtesy of ORNL.

band of equivalent method results in reversing the transformer design
choice. Using the band of equivalent approach effectively reduces the
value of the loss evaluation factors. It usually results in picking the
less efficient transformer. Typically, the band of equivalent of total
owning cost is 1 to 3 percent. In the example in Table 6-1, the cut-over
point for changing from selecting design 1 as the most cost-effective
choice to design 2 as the most cost-effective choice can be calculated by
using the following formula:

$$\$490 + x(4.5)(100) + x(1.00)(220) = \$475 + x(4.5)(80) + x(1.00)(280)$$

Solving for x,

$$x = \frac{15}{30} = 0.5$$

where x is the reduction in cost of no-load and load losses at which
design 2 instead of design 1 becomes the most cost-effective transformer. In this case, a 3 percent ($1100/1075 = 1.03$) band of equivalent
approach reduces the cost of no-load losses by 50 percent from
$4.50/W to $2.25/W and the cost of load losses by 50 percent from
$1/W to $0.50/W. It is seen clearly in this example that the band of
equivalent approach does not really deal with uncertainty but instead
reduces the cost of losses and results in picking a less efficient transformer. According to the Oak Ridge National Laboratory report entitled, *Determination Analysis of Energy Conservation Standards for
Distribution Transformers,* "There is a need to develop a better method
(other than band of equivalent) to incorporate uncertainty into the
total owning cost selection process. A minimum efficiency criterion

may be one of the ways to promote purchases of more efficient transformers." This is why the EPA's Star Transformer Program of setting minimum efficiency values for various sized transformers has great value.

There is still the need to deal with the uncertainty of load growth, cost of losses, interest rates, and inflation. This is especially true when the utility industry is being deregulated. The effect of deregulation is uncertain and needs to be dealt with in a systematic way. The use of sensitivity analysis is one way to deal with uncertainty.

Sensitivity Analysis

Sensitivity analysis is a method for determining the effect of changes in the components of the total owning cost formula on the overall total owning cost results. It is accomplished by assuming small changes in these components and calculating the resulting change in the total owning cost. These components include fixed charge rate, minimum rate of return, system energy and capacity costs, transformer loading, magnitude of no-load and load losses, and projected inflation. The price of the transformer is usually based on bid prices and needs to be evaluated for its sensitivity to possible change.

The first step in performing a sensitivity analysis is to develop a base case of total owning cost based on the most likely total owning cost component values. The next step is to vary the costs of the total owning cost components from the base case. Then the changes in total owning cost can be plotted on a graph as one total owning cost component changes while the others remain the same. Once the point at which a particular total owning cost component incremental change results in a change in the decision as to what transformer design is cost-effective, the decision maker can decide if this change is likely to occur.

The 1963 and 1981 utility industry studies reports both contained sensitivity analyses (see Table 6-2). From these analyses it was determined that certain total owning cost components required large changes to affect the total owning cost results by more than 1 percent. These components included the cost of fixed capacitors, cost of switched capacitors, price of energy, responsibility factor, load factor, and load change caused by a change in the voltage. It can be said that the total owning cost method is relatively insensitive to these components.

One method of performing a sensitivity analysis is to calculate the parameter sensitivity. *Parameter sensitivity* is defined as the percentage of input parameter variation required for a 1 percent change in the total levelized cost of ownership output parameter. Large numbers are

Table 6-2. Sensitivity Comparison of 1963 Utility Industry Studies 1981 EEI Report

Input parameters	Parameter sensitivity (percent)	
	1963 study	1981 study
Fixed charge rate (FC)	2.5	2.4
Cost of system capacity (SC)	0.7	0.7
Cost of energy (EC)	1.2	1.1
Cost of fixed capacitors	1098	NE
Cost of switched capacitors	412	NE
Sale price of incremental energy	13.3	NE
Equivalent annual peak load (PL)	1.3	1.2
Loss factor of transformer load (LF)	12.8	12.0
Coincidence factor of transformer Load to system peak load (RF)	3.8	3.6
Coincidence factor of transformer Loss to feeder peak load	813	NE
Load power factor	21	NE
Percent change load per percent Change voltage	20	NE
Winding loss (LL)	3.5	3.3
Excitation loss (EL)	3.9	3.7
Bid price (BP)	2.5	2.4

an indication that the total owning cost is insensitive to change in parameters. The parameter sensitivity ratio can be determined from the following formula:

$$\text{Parameter sensitivity ratio in percent} = \frac{\text{percent change input} \times 10^2}{\text{percent change output}} \quad (6.1)$$

Table 6-3, from the 1981 study, shows the parameter sensitivity ratio for the various parameters in the total owning cost formula. This table shows that the most sensitive component in the total owning cost formula is the system capacity cost. It also shows that the least sensitive component is the transformer loss factor. For this analysis, it is helpful to identify the range of system capacity expected by the transformer purchaser. Graphing the expected value of the more sensitive components, such as system capacity, against total owning cost is the best way of dealing with uncertainty. Then the transformer purchaser can decide how these more sensitive components will affect his or her decision.

There are other factors that need to be considered in evaluating the total owning cost of a transformer. These include the environmental

Table 6-3. Sensitivity Parameter Analysis of 1981 Study

Input parameters	Parameter sensitivity (percent)	
	Input variation	Output variation
Fixed charge rate (FC)	± 5 percent	± 2.1 percent
Cost of system capacity (SC)	± 5 percent	± 7.1 percent
Cost of energy (EC)	± 5 percent	± 4.5 percent
Equivalent annual peak load (PL)	± 5 percent	± 4.2 percent
Loss factor of transformer load LF)	± 5 percent	± 0.4 percent
Coincidence factor of transformer Load to system peak load (RF)	± 5 percent	± 1.4 percent
Load loss (LL)	± 5 percent	± 1.5 percent
No-load loss (NLL)	± 5 percent	± 1.4 percent
Bid price (BP)	± 5 percent	± 2.1 percent

effects of transformer losses, transformer reliability, and the effect of operating temperature on total owning cost. Each one of these items will be dealt with separately.

Environmental Economics

Environmental economics can be another parameter that affects the decision to purchase an energy-efficient transformer. Efficient transformers have less loss and require less generation to deliver that loss. The less generation results in fewer emissions into the air. These environmental effects of electric utility generation emissions into the air have become more of a concern by both federal and state governments. This concern is best described in the 1992 Energy Policy Act; Subtitle B, Utilities, Section III states that "rates charged by any utility shall be such that the utility is encouraged to make investments in, and expenditures for, all cost-effective improvements in the energy efficiency of power generation, transmission, and distribution...." Consequently, state regulators are requiring utilities to account for the environmental costs of their decisions. How to deal with the cost of air pollutants from incremental electrical generation was the subject of a 1994 report by General Electric entitled, *Guide for Evaluation of Distribution Transformers*. This report refers to the E factor, "which would provide an evaluation of the impact of environmental externalities." The report states that "the E factor for a 25-kVA transformer with core loss of 30 watts, an assumed levelized fuel cost of externalities of

2 ¢/kWh (generation to supply the losses), and a fixed charge rate of 17 percent would be

$$(0.30) \times (8760) \times (0.02)/0.17 = \$30.92$$

Thus, the additional cost to be added to the total owning cost is $30.92."

Even though the environmental cost of transformer losses is not a direct cost to the transformer user, it is certainly a cost to society. Depending on how environmentally sensitive the transformer user wishes to be in its evaluation of transformer losses, this approach gives a method for taking the environmental cost into account. This method involves adding to the A and B factors the cost of no-load and load losses associated with emissions. The cost depends on the type of emissions associated with a specific type of generation. It requires determining the cost of emissions and the amount of emissions associated with a kilowatthour of energy production. It may be an added cost that a transformer user should consider in its overall evaluation of transformer losses.

Reliability Economics

Another factor to consider in deciding what transformer to buy is reliability. More and more utilities are including the cost of reliability along with the costs of the transformer and losses in evaluating different transformer designs and manufacturers. This means knowing the failure rate history of various design and manufacturing processes. Based on this historical database, many utilities have developed various methodologies for evaluating the reliability of transformers. They are also finding in general that high-efficiency transformers are more reliable. The evaluation of reliability provides an incentive to manufacturers to build more reliable transformers. It has resulted in reduced failure rates and better communication between transformer purchasers and suppliers. Each reliability evaluation methodology includes a four-step process: (1) develop a database of failures, (2) determine the failure rate of various designs and manufacturers, and (3) apply a cost to the failure rates, and (4) include the cost of failures in the total owning cost formula.

In order to include reliability in the total owning cost formula, it is first necessary to develop a database that contains the number of transformers purchased and failures categorized by manufacturer's name, manufacture date, failure description and date, serial number, type, size, and voltage. In a 1991 *Electrical World* special report entitled,

"How to Buy the Best Transformer," John Reason says that "manufacturers must have 200 service years on the system before an individual failure rate is calculated. Manufacturers with fewer than 200 service years are assigned the system's average failure rate." Many utilities record the data on transformer failures in a computerized record-keeping system.

The next step is to determine the hazard and failure rate. The *hazard rate* is the number of transformer failures in a year as a percentage of the number of transformers in service at the beginning of that year. The hazard rate can be calculated using the following formula:

$$\text{Hazard rate} = \frac{\text{number of failures in year}}{\text{number of units in service at the beginning of year}} \quad (6.2)$$

The *failure rate* is defined as the total number of failures as a percentage of the total unit-years of service for a batch of transformers bought from a certain manufacturer at one time. Therefore, the failure rate can be calculated using the following formula:

$$\text{Failure rate} = \frac{\text{total number of failures}}{\text{total unit-years of service}} \quad (6.3)$$

The resulting hazard and failure rates can be graphed. Figure 6-3 illustrates the familiar "bath tub" curve showing the relationship between years of service and failure rate.

Based on the described failure-rate analysis, a utility can determine the corresponding cost of the failures. Wisconsin Public Service (WPS) Corporation began in the 1980s developing failure-rate costs utilizing a

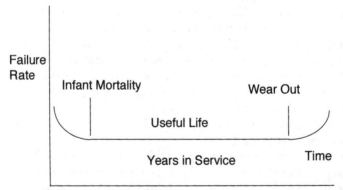

Figure 6-3. Years of service versus failure rate. (*Courtesy of Wisconsin Public Service Corp.*)

failure-rate computer program developed by Don Duckett of General Electric. In a 1991 *Transmission and Distribution* article entitled, "Failure Analysis Improves Distribution Transformer Quality," Michael Radtke of WPS showed failure-rate costs for 10-kVA overhead transformers and single-phase pad mounts purchased between 1987 and 1990. Figures 6-4 and 6-5 show these failure costs. WPS has provided the failure-rate costs to various transformer manufacturers to provide them with an incentive to reduce their failure rate. WPS has experi-

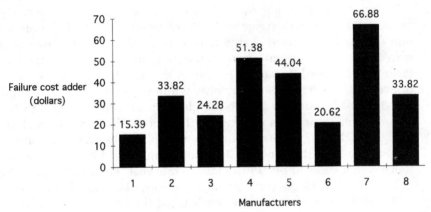

Figure 6-4. Failure-rate costs for 10-kVA overhead transformers. (*Courtesy of Wisconsin Public Service Corp.*)

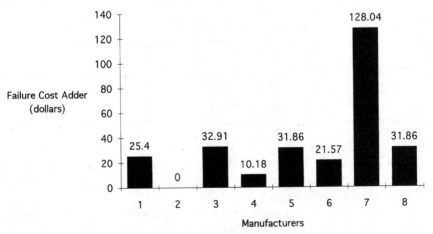

Figure 6-5. Failure-rate costs for 25-kVA padmounted transformers. (*Courtesy of Wisconsin Public Service Corp.*)

enced a 30 to 50 percent failure-rate reduction since this program was introduced in 1983.

The failure-rate cost includes three components: (1) the labor cost of replacing the failed transformer with a repaired or new transformer, (2) the capital cost including carrying charges of either repairing or replacing the damaged transformer, and (3) the transformer salvage value. The labor cost is treated as an expense and is drawn directly from revenues, while the capital cost and salvage values are lumped together and treated as a capital investment. The labor cost necessary to change out the transformer is treated as an expense and comes from revenues rather than borrowed funds. It is necessary to convert the capital cost of the new transformer into a levelized annual amount by multiplying by the fixed charge rate. This levelized annual amount is converted into a present-worth value by multiplying the series uniform present-worth factor for the life of the transformer minus 1 year to account for the 1-year warranty, while the labor cost is a single amount and is multiplied by the single payment present-worth (*SPW*) factor to convert it into a present-worth value. Both the capital replacement and labor costs are escalated by multiplying by an escalation factor. In each case, the annual failure rate is multiplied by the labor and replacement costs to obtain the failure-rate cost. Before including the failure-rate cost in the total owning cost formula, it needs to be converted to an equivalent first-cost basis using the following formula:

$$
\begin{aligned}
\text{Failure-rate cost} = {} & \text{FR} \times \text{USPW} \times \text{FCR} \times \text{replacement price} \\
& \times \text{escalation factor} + \text{FR} \times \text{cost of change out} \\
& \times \text{SPW} \times \text{escalation factor} \qquad\qquad (6.4)
\end{aligned}
$$

where
FR = annual failure rate
$USPW$ = uniform series present-worth factor
FCR = fixed charge rate
Replacement price = replacement price of new or repaired transformer
Escalation factor = inflation rate
SPW = single present-worth factor
Cost of change out = the annual labor cost to replace damaged transformer

This formula can best be understood by an example. Table 6-4 contains the parameters for calculating the reliability cost or failure-rate cost of two-transformer designs.

Based on these parameters and the preceding formula, the failure-rate cost for design 1 with a failure rate of 1.3 percent and design 2 with a failure rate of 0.5 percent can be shown in Table 6-5. As can be

Table 6-4. Cost of Reliability Evaluation

Transformer values	Design 1	Design 2
Transformer price	560	600
Annual failure rate	1.3%	0.5%
Replacement labor cost	$300	$300
Replacement cost	$560	$600
Escalation rate	5%	5%
Fixed charge rate	17%	17%
Cost of capital	12%	12%
Economic life	25 years	25 years

SOURCE: 1990 IEEE paper by Daniel Wood entitled, "Evaluating Product Reliability Costs."

Table 6-5. Cost of Reliability

Transformer costs	Design 1	Design 2
Replacement capital	$ 60.32	$ 24.86
Replacement labor	33.46	12.87
Initial price	560.00	600.00
Total	$653.78	$637.73

SOURCE: 1990 IEEE paper by Daniel Wood entitled, "Evaluating Product Reliability Costs."

seen from this table, the lower-cost design 1 has a higher reliability cost, which makes it more costly than the higher-cost design 2 with the lower reliability cost. These types of calculations can be made with commercially available software by companies such as General Electric.

Finally, once the failure-rate cost is determined by transformer size, voltage, and manufacture, this value is added to the total owning cost formula:

$$TOC = \text{initial cost} + \text{failure cost} + \text{no-load loss cost} + \text{load loss cost} \quad (6.4)$$

Energy-efficient transformers tend to be more reliable. This is so because energy-efficient transformers have less loss and internal heating. The lower heating reduces the deterioration of the insulation and extends the life of the transformer.

Transformer Evaluation
Computer Programs

Several transformer manufacturers have developed computer programs for calculating total owning cost. In addition to the programs developed by the transformer manufacturers, the EPA has developed a computer program for evaluating transformer losses. (The EPA computer program is included on a disk, which is on the inside back cover of the book.) These programs are all similar in their approach in that they use the total owning cost method. They are usually free and will run on Windows. Some transformer manufacturers that have developed these computer programs include ABB, GE, and Cooper. They can be obtained by contacting the local sales representative of one of the transformer manufacturers.

7
Transformer Replacement

Transformer replacement can be caused by a failure, over- or under-loading, distribution line upgrade, or the need to replace existing low-efficiency transformers with new higher-efficiency transformers. In any case, the effect would be significant if all the distribution transformers were replaced with energy-efficient transformers. An April 1995 report by Oak Ridge National Laboratory entitled, *The Feasibility of Replacing or Upgrading Utility Distribution Transformers During Routine Maintenance* (ORNL-6804), stated that "...on the order of 44 billion kWh annually could be saved if all distribution transformers were immediately replaced by new low-loss units." However, the report goes on to say that this is not practical because most of these replacements would not be cost-effective. Although the report does say that from an analysis of 60 investor-owned utilities that operate a third of the transformer capacity in the United States, "approximately 87 percent of transformer refurbishment is economically justified." Whether a transformer is replaced as part of routine maintenance or as part of a planned replacement program, there are four options for replacing the transformer.

1. Reinsulation
2. Rewinding
3. Refurbishment
4. New transformer

The economics of early transformer replacement is determined by comparing the life-cycle cost of each option. In the life-cycle cost analysis, the present values for capital and energy are calculated by taking the present value of the capitalized total owning cost. Comparing transformer costs using present value or capitalized costs does not change the comparison, although the absolute values would be different. The life-cycle cost of a new transformer is straightforward. The first step in making the comparison is to determine the remaining life of the existing transformer. The next step is to assume that the existing transformer will be replaced by the same transformer now or later. The total life-cycle costs for each alternative need to be converted to present values. It is assumed that only transformers taken out of service for routine maintenance are candidates for replacement. Therefore, the costs of reinstalling the transformer are the same for both alternatives and are not included in the analysis.

The total owning cost formula is used to determine which option is the most cost-effective. How to determine the most cost-effective option is the subject of this chapter. It involves weighing the costs of the option against the energy savings and increased life of the option. In evaluating whether or not to replace an existing transformer, the two most important factors are the cost to repair or replace the existing transformer and the assumed remaining life. Where there is uncertainty as to the remaining life or cost of replacement, a sensitivity analysis is appropriate. The flowchart in Fig. 7-1 illustrates the process of deciding which option is the most cost-effective.

An aggressive program to replace older transformers with newer energy-efficient units is worth considering. Many utilities develop an age criterion that triggers replacement. Based on an Oak Ridge National Laboratories survey, 49 percent of utilities had an age criterion that varied from 14 to 35 years, with an average of 25 years. The age criterion depends on the value of losses and the amount of losses assumed for older and newer transformers. Significant savings in losses and money may be obtained by replacing an older unit with a newer energy-efficient transformer.

Routine maintenance can result in a decision to replace or repair an existing transformer. Many utilities have an inspection requirement that gives the maintenance crew some guidelines on whether a unit needs to be removed for repair or replacement. These guidelines are based on inspection tests that may include visual inspection for obvious problems, insulation power factor tests, tests of the oil or presence of polychlorinated biphenyls, turns ratio tests, voltage and current tests, and loss tests.

If the unit is to be replaced rather than repaired, the scrap value of the old unit needs to be included in the total owning cost formula. The

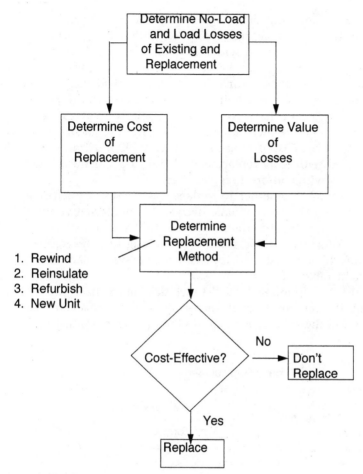

Figure 7-1. Transformer replacement economics method.

scrap value is subtracted from the cost of a new unit. The scrap value depends on the condition of the older unit and whether the unit can be sold and recycled. This involves taking the total owning cost formula and modifying it to reflect the life-cycle cost of replacing a transformer:

$TOC_{replacement}$ = replacement cost − salvage value
+ cost of no-load losses of replacement + cost of load losses of replacement

The preceding total owning cost of replacement must be compared with the total owning cost of continuing to operate the existing transformer. The cost of replacement depends on whether the unit is to be replaced with a new unit or is to be refurbished. If the unit is to be refurbished, there is no savings associated with salvage, while if the existing unit is to be replaced with a new transformer, there is the salvage value of the existing unit that needs to be taken into account. The decision as to whether to return the existing transformer to service with no change, to replace it with a new unit, to rewind it, to reinsulate it, or to refurbish it is an economic decision. It involves weighing the increased life and reduced losses resulting from replacing the existing unit against the cost of replacing it. Figure 7-2 illustrates the cash flow associated with early replacement. Figure 7-3 illustrates the cash flow associated with end-of-life replacement.

The decision as to whether to replace the existing transformer or return it to service is an economic decision. The alternative with the lowest total owning cost is the most cost-effective alternative. In the case of a transformer with 10 years of remaining life, the comparison of the timing to refurbish and replace later versus replacement now can be seen in Table 7-1.

The "first now" option in Table 7-1 includes the levelized capital and energy costs throughout the analysis period. The "refurbish and replace later" option in the same table includes the following: (1) in year 1 the

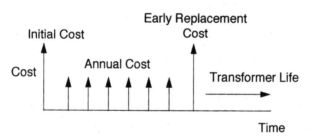

Figure 7-2. Early replacement cash flow diagram.

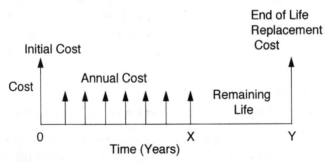

Figure 7-3. End-of-life replacement cash flow diagram.

Table 7-1. Transformer Early Replacement versus
Refurbishment Timing

| Year | Option and costs | |
	Refurbish and replace later	Replace now
1	Refurbishment costs plus levelized energy for refurbished transformer	Levelized energy and capital for new transformer
2–9	Levelized energy for refurbished transformer	Levelized energy and capital for new transformer
10	Takedown and reinstall costs plus levelized energy and capital for new transformer	Levelized energy and capital for new transformer
11–30	Levelized energy and capital for new transformer	Levelized energy and capital for new transformer

SOURCE: Courtesy of ORNL.

refurbished transformer levelized energy and capital costs, (2) in years
2 through 9 the new transformer takedown and reinstallation, (3) in
year 10 the new transformer levelized capital and energy costs, and (4)
in years 11 through 30 the levelized capital and energy costs. The three
cost components of the two alternatives are capital, energy, and change-
out and refurbishment costs. The "refurbish and replace later" option
present value is less than the "replace now" option when the capitalized cost of the new transformer can be sufficiently delayed, while the
"replace now" option has the advantage of not including the extra costs
of refurbishment, takedown, and reinstallation. The "replace now"
option eliminates the refurbishment, takedown, and reinstalling costs
in year 10. The other difference between the two alternatives is the difference in losses in years 1 to 10, while the losses for the two alternatives are the same for years 11 to 30. The cost-effectiveness of the two
alternatives is a weighing of the value of energy efficiency against the
remaining life of the refurbished transformer.

Using the time profile from Table 7-1, the following example, taken
from the ORNL-6804 study, shows how to calculate the total owning
cost of replacement versus refurbishment for a 50-kVA transformer.

Example Assuming that the cost of the new transformer is $734 and
that the cost to changeout and refurbish the existing transformer is $201,
and assuming that the life of a transformer is 30 years and that the remaining life of the existing transformer is 10 years, should the utility replace
the existing transformer with a new one or refurbish the old transformer
and return it to service?

solution   The present worth of the total owning cost of the two alternatives is calculated as follows:

Replacement Alternative:

Present worth of $TOC_{\text{replacement}}$

$$= \text{cost of replacement} - \text{salvage value}$$
$$+ \, PW \text{ of cost of no-load losses of replacement}$$
$$+ \, PW \text{ of costs of load losses of replacement}$$

where   Cost of replacement = $734
Salvage value = $0
Present worth of no-load losses of replacement = $212
Present worth of load losses of replacement = $673

Therefore

Present worth of $TOC_{\text{replacement}}$ = $734 − 0 + $212 + $673 = $1619

Refurbishment Alternative:

Present worth of $TOC_{\text{refurbishment}}$

$$= \text{cost of refurbishment}$$
$$+ \, PW \text{ of cost of no-load losses of refurbished transformer}$$
$$+ \, PW \text{ of cost of load losses of refurbished transformer}$$
$$+ \, PW \text{ of cost of replacement in the 10th year}$$

where   Cost of refurbishment = $201
Present worth of no-load losses of refurbishment = $378
Present worth of load losses of refurbishment = $628
Present worth of replacement in the 10th year = $373

Therefore

Present worth of $TOC_{\text{refurbishment}}$ = $201 + $378 + $628 + $373 = $1580

Based on this example, it is more cost-effective to refurbish the existing transformer rather than replace it with a new unit.

Replacement Options

When a transformer is removed from service during routine maintenance, the utility or industrial facility has to decide what to do with the existing unit. Should it rewind, reinsulate, refurbish, or replace the existing transformer? There are three factors that determine this decision: age, transformer size, and future load growth. If a transformer is too old or too small, it may not be feasible to refurbish it. If additional load growth is expected, a new, larger-capacity transformer may be a better choice than trying to reuse the existing unit. Each one of these replacement options needs to be examined before making this decision.

Rewinding

Rewinding a transformer can result in increased life and reduced losses. If the core and tank are in good condition, they are usually left intact. If the core is not replaced, the no-load losses usually increase due to handling of the core. If the core is replaced with low-loss silicon steel, however, the no-load losses may go down, while the load losses are reduced if the transformer winding is replaced with larger conductors or in the case of aluminum conductors with copper.

The decision as to whether to rewind or replace a transformer hinges on the cost versus the change in the losses. The cost of a rewind can be significantly less than the cost of replacing an existing transformer with a new one. Oak Ridge National Laboratories (ORNL) found in a recent study that for 25-kVA transformers, "rewinding cost was about 76 percent of the average cost for a new transformer." Often utilities find that it is not cost-effective to rewind transformers of advanced age, such as 30 years. This is so because the various components, such as insulation and core, of the transformer have deteriorated due to aging. Often the accessories, such as bushings, tap changers, gauges, radiators, and other cooling equipment, cannot be repaired or replaced because parts are no longer available.

Most utilities do not rewind existing transformers to extend their life. The ORNL survey showed that "rewinds constitute less than 2 percent of refurbished transformers." This is so because of the uncertainty of reliability and losses associated with rewound transformers. The energy savings are usually not significant because rewinding does not usually reduce the no-load losses.

Reinsulation

Transformer reinsulation can result in longer life, increased capacity, and lower losses. Many utilities are using a new type of insulation material to reinsulate failed or damaged transformers. This new insulation material is called *aramid*.

Aramid is a manufactured thermoplastic aromatic polyamide that can withstand much higher temperatures than conventional transformer insulating paper and pressboard. Conventional insulation is limited to an average winding temperature of 65°C above an ambient temperature of 40°C, while aramid insulation can withstand 200°C continuously. Consequently, a transformer reinsulated with aramid insulation can increase its load-carrying capability significantly. For example, a Florida utility recently increased a 100-MVA transformer capacity to 120 MVA by reinsulating with aramid insulation. The rating increases of the three

Table 7-2. Transformer Ratings Before and
After Reinsulation

| | | Ratings (MVA) | |
Winding	Original	Conventional insulation	Aramid insulation
High voltage	60/80/100	60/80/100	72/96/120
Low voltage	39/52/65	60/80/100	72/96/120
Tertiary	21/28/35	21/28/35	21/28/35
Total cost	Unknown*	$520,000	$690,000

*Installed in 1978.
SOURCE: T&D Special Report, *Electrical World,* March 1991, pp. 10–11.

windings of this transformer are described in Table 7-2. As can be seen from the table, the primary disadvantage to reinsulating with aramid insulation is the extra cost, although the cost of reinsulating with aramid insulation is half the cost of a new transformer.

A utility can decide to operate the reinsulated transformer at the same rating with lower losses or at an increased rating with the same losses. The transformer load losses can be reduced by increasing the cross-sectional area of the copper conductor without increasing the size of the transformer. The dielectric properties of aramid insulation permit reduced clearances and allow a larger conductor without increasing the size of the core space. For example, one California utility found that by reinsulating a failed 18/24/30 MVA OA/FA/FA transformer with aramid insulation it could reduce total losses up to 14 percent and load losses by 21 percent, while a Virginia utility was able to reduce the losses of a 12-MVA transformer by 16 percent at base rating by reinsulating with aramid insulation. Figure 7-4, obtained from a November 1989 *Electrical World* article entitled, "New Insulation Improves Transformer Performance," illustrates loss reduction versus loading for a transformer reinsulated with aramid insulation.

In order to keep the cost of reinsulation down, the more expensive aramid insulation is limited to the hottest areas of the transformer. According to the November 1989 *Electrical World* article, for "lower load losses or a minor uprate, 20 to 30 percent aramid insulation is necessary; more aramid, 30 to 50 percent, is needed for a major uprate."

When reinsulating with aramid, the oil becomes the temperature-limiting factor rather than the insulation. Oil temperature can be kept within limits by the addition of cooling equipment, i.e., fans, pumps, and radiators. Also, the copper cross-sectional area can be increased to reduce losses and increase capacity. Another option to replacing an existing transformer is to refurbish the existing transformer.

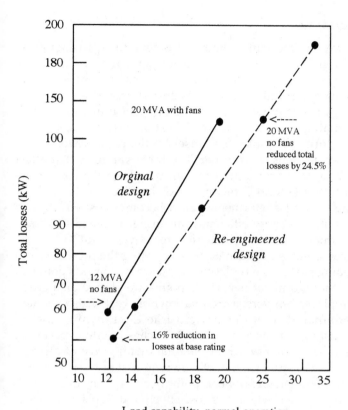

Figure 7-4. Loss reduction versus loading with aramid insulation. (*Courtesy of Electrical World.*)

Refurbishment

Refurbishment involves a minimal replacing of transformer parts, such as connectors, nuts, gaskets, and bushings. It may involve changing the transformer oil or drying out the coils.

According to the ORNL-6804 report, "...about 1.0 percent of the total installed transformer capacity of utility distribution transformers was refurbished in 1993. This would be about 17 million kVA for the entire nation. Most utilities did not have readily available information on the age of their refurbished transformers. Based on those reporting, about 47 percent of refurbishment activity occurs on transformers that are less than 10 years old, and about 81 percent occurs on transformers less than 20 years old. Only about 1 percent of total refurbishment's (by kVA) occurs on transformers that are 30 years old or older."

Sensitivity Analysis

One effective way to determine whether it is time to replace an existing transformer with a more efficient transformer is to develop an age criteria based on sensitivity analysis. Sensitivity analysis or break-even analysis involves determining when it is cost-effective to replace an existing high-loss transformer with a more efficient transformer. This involves determining if the present-worth cost of early replacement of an existing transformer is more than offset by the present-worth savings of an energy-efficient transformer. The three variables that affect this determination vary from utility to utility and are the cost of no-load and load losses (A and B factors) and the discount factor. The point in the life of the transformer when it becomes cost-effective to replace it with a more energy-efficient transformer is called the *break-even point*. How does a utility calculate the break-even point?

The break-even point is determined by calculating the present worth of the total owning cost of the replacement assuming various points in the life of the existing transformer. These points are plotted on a graph of the year of replacement versus the total owning cost. This exponential curve is compared with the straight-line total owning cost of replacing the existing transformer when it reaches its life expectancy. The point when the present worth of the total owning cost of replacement curve crosses the straight-line total owning cost of replace later is the break-even point. The break-even point becomes the time in that particular utility's system when it is cost-effective to replace an existing transformer with an energy-efficient transformer. This provides the utility an age criterion on whether to replace or refurbish a transformer it removes from service for routine maintenance.

Based on a survey of 60 investor-owned utilities, Oak Ridge National Laboratories performed a break-even analysis on variously sized distribution transformers. Using the average costs and losses from this survey, ORNL produced the results shown in Table 7-3. This table can be used as a benchmark of break-even transformer lives to determine whether it is cost-effective to replace or refurbish an existing transformer. If the remaining life of the existing transformer is less than the break-even life, then it is cost-effective to replace the existing transformer. If the remaining life of the existing transformer is greater than the break-even life, then it is not cost-effective to replace the existing transformer. In order to determine the transformer break-even age, the utility must first assume the life of the new distribution transformer and the corresponding life of the existing transformer based on when it was installed on the system. The break-even age becomes an age criterion for determining when it is cost-effective to replace or refurbish an existing transformer.

Table 7.3. Transformer Replacement and Refurbishment Cost

Transformer type and size (kVA)	Average cost of new transformer	Average no-load loss (W)	Average load loss (W)	Average cost to refurbish	Average cost of exchange	Life after break-even point	Age at break-even point
Pole							
10	$ 396	31	151	$ 130	$ 339	14	21
15	$ 450	40	212	$ 137	$ 370	13	21
25	$ 543	58	312	$ 168	$ 365	13	22
37.5	$ 671	81	412	$ 168	$ 422	12	23
50	$ 777	99	520	$ 220	$ 409	12	23
Pad type							
50	$ 1,129	98	536	$ 309	$ 533	10	25
75	$ 1,406	133	718	$ 309	$ 533	9	29
167	$ 2,264	256	1350	$ 314	$ 623	7	32
Three-phase							
225	$ 4,892	396	1998	$ 755	$ 926	10	25
500	$ 7,197	721	4021	$1346	$1410	14	20
1000	$11,503	1230	7246	$1346	$1410	10	25
Average age						11.3	24.2

SOURCE: Courtesy of ORNL.

In order to calculate the total owning cost of keeping an older transformer and refurbishing it, the utility must know the no-load and load losses of the existing unit. This is often difficult to determine. The losses for existing transformers of various vintages were obtained from the ORNL-6804 report and are listed in Tables 7-4 and 7-5.

The assumed life of the existing transformer is critical to the whole analysis. A shorter life has the effect of reducing the value of the savings and increasing the value of the takedown and reinstallation costs. The assumed life is very important because of its uncertainty and effect on the results of the comparison.

Another important assumption is the effect of the discount rate on future costs. The discount rate deals with the time value of money and reduces the effect of future costs. A lower discount rate favors keeping the refurbished

Table 7-4. No-Load Losses for Distribution Transformers

Year	Transformer size (kVA)					
	15	25	37.5	50	75	100
1960	87	127	171	212	287	357
1965	81	118	159	197	267	331
1970	79	115	155	192	260	323
1975	71	103	139	172	233	289
1980	62	90	123	152	205	255
1985	54	78	107	131	177	221
1993	40	58	81	99	133	166

SOURCE: Courtesy of ORNL.

Table 7-5. Load Losses for Distribution Transformers

Year	Transformer size (kVA)					
	15	25	37.5	50	75	100
1960	234	343	465	578	783	972
1965	211	309	419	521	706	876
1970	191	279	378	470	637	790
1975	196	286	385	481	655	809
1980	200	293	393	492	672	827
1985	205	301	400	503	690	845
1993	212	312	412	520	718	875

SOURCE: Courtesy of ORNL.

transformer longer due to the reduction in the value of loss savings. A lower discount rate increases the importance of the future changeout replacement costs. The federal Office of Management and Budget (OMB) requires a 7 percent real discount rate. The discount rate, present worth, and other engineering economic terms are defined in the appendices.

New Transformer

A new transformer has the decided advantage of having lower losses than an existing older transformer. There are some other advantages as well. A new transformer can be sized to fit the load, while the older existing transformer may no longer match the load. A new transformer has a warranty, while the older transformer's warranty has expired. If the older transformer contains PCBs, it should be replaced with a newer transformer. And finally, there is the expected longer life and higher reliability of a new transformer. Other than evaluating the economics of a new transformer versus refurbishing the existing unit, how does a utility decide whether to replace an older transformer with a new one? The key to making the best decision is a good maintenance testing program.

Maintenance

Good maintenance practices will ensure that existing transformers operate efficiently and reliably. These practices also will provide the necessary data to help determine whether an older transformer should be replaced with a newer transformer. Inspections and tests of the transformer oil, gaskets, winding insulation, radiators, and electrical grounds are recommended. The maintenance program of most utilities includes inspection, testing, minor and major refurbishments, and retirements. Figure 7-5 is a flowchart based on the Oak Ridge National Laboratories survey of 60 utilities that illustrates the percentages of distribution transformers that are refurbished, rewound, and replaced as a result of routine maintenance.

Oil

Regular testing of the condition of the transformer oil is essential to a good maintenance program. The condition of the oil affects not only the insulation but also the cooling. It is essential that the oil be free of contaminants such as sulfur, moisture, and any other corrosive elements. Tests have been developed for the purpose of detecting these contaminants.

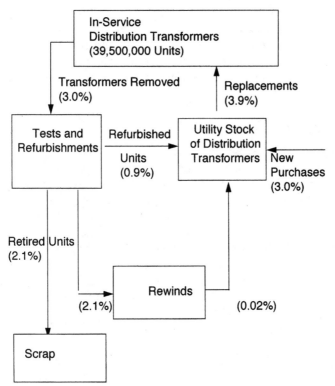

Figure 7-5. Distribution transformer routine maintenance.
(*Courtesy of ORNL.*)

Two simple field tests will provide the status of the oil. One test is subjective and involves comparing the color of the transformer oil to a series of colors to see if there is any deviation of the color from the standard. As oil ages and deteriorates, it becomes darker and discolored. The other test is a more objective test that involves testing for the breakdown of voltage and moisture content in the oil. A voltage is applied to a sample of the oil until a breakdown happens. The oil is unacceptable if the breakdown occurs below 20,000 V. If the water content of the oil is above 30 parts per million (ppm), the oil should be processed per the manufacturer's instructions. For more information on these oils tests, see ANSI/IEEE C57.106, *Guide for Acceptance and Maintenance of Insulating Oil in Equipment.*

Gases

Inspection of the transformer should include testing for gases in the oil or nitrogen blanket. A sample of the oil brought to the laboratory can be

Table 7-6. Combustible Gas Limits

Combustible gas	Concern limit (ppm)
Hydrogen (H$_2$)	100
Methane (CH$_4$)	120
Acetylene (C$_2$H$_2$)	035
Ethylene (C$_2$H$_4$)	050
Ethane (C$_2$H$_6$)	065
Carbon monoxide (CO)	350

SOURCE: ANSI/IEEE C57.104, *Guide for Interpretation of Gases Generated in Oil-Immersed Transformers.*

tested for dissolved gases by the use of a gas chromatograph. IEEE has developed a guide for determining acceptable levels of gases in transformer oil. The guide is entitled, *Guide for Interpretation of Gases Generated in Oil-Immersed Transformers* (ANSI/IEEE C57.104). Table 7-6 provides a breakdown of these gas levels.

Winding Insulation

The major tests for determining winding insulation condition are the "Meggar" test, winding ratio test, insulation power factor test, the low-voltage impulse test, and the low voltage excitation current test.

The "Meggar" test involves applying a dc voltage to the winding and measuring with a megaohmeter the resistance between windings at the terminals and the ground. The measured values at different temperatures are compared with specified values to see if there are any deviations from the standards specified in ANSI/IEEE C57.125.

The winding ratio can be measured by applying a 120-V ac current to the primary winding and measuring the voltage induced in the secondary winding. The winding ratio should be within 0.5 percent of the ratio of rated voltages. Any deviation from this tolerance may indicate shorted turns or an open winding.

The insulation power factor test measures the resistive current by measuring the power factor of the insulation when an ac voltage is applied to the transformer winding. A low power factor indicates a low-resistive current and thus satisfactory insulation. A power factor above 1 percent indicates deteriorating insulation that needs further investigation.

The impedance test is conducted by applying a voltage to the primary winding and measuring the current in the secondary winding. The ratio of voltage to current should be within 2 percent for each phase and should not vary by more than 2 percent between tests. Any variation greater than 2 percent is an indication of possible winding damage. This test is described in detail in ANSI/IEEE C57.125.

8

High-Efficiency Transformers

High-efficiency transformers are available commercially today. The transformer manufacturers are able to provide myriad choices of transformers of varying losses. It all depends on how high the transformer purchaser values losses. As utilities and other large purchasers have insisted on lower-loss high-efficiency transformers, the transformer manufacturers have been able to develop transformers with lower losses.

The reduction in losses over the years have come by improving the materials and construction of the core and coils of distribution and substation transformers. With the advent of low-loss amorphous steel core transformers in the 1980s (see Chap. 9) and widespread use of evaluation methodologies (see Chap. 6), a steady reduction in no-load and load losses has occurred in the transformer industry. These reductions have resulted in the development of low-loss silicon steel core transformers and advanced conductor design.

Advanced Conductor Design

Conductor design has been improving over the last 20 years to reduce transformer load losses. Copper conductors have replaced aluminum conductors due to their lower resistance and higher tensile strength. The transformer industry has been looking at advanced conductor design in order to improve the efficiency of transformers. One of the latest developments in conductor design is in the use of ribbon conductors.

The ribbon conductor reduces load losses by reducing the eddy losses and allowing more copper to be installed in a smaller space. The rib-

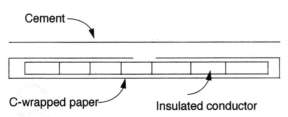

Figure 8-1. Ribbon conductor construction. (*Courtesy of Electrical World.*)

bon conductor reduces the eddy current losses by reducing the cross-sectional dimension of the conductor. A rule of thumb is that for every percentage reduction in cross-sectional dimension, the eddy current losses are reduced 1½ percent. This also improves the space factor by over 25 percent. It is constructed of seven segments of flat subconductors cemented edge to edge (Fig. 8-1). Axial cooling in the conductor is required to keep the temperature in the conductor within standards.

In order to test the loss-reduction features of the advanced ribbon conductor, Houston Lighting and Power Company (HL&P) in 1991 obtained two ABB-manufactured 25/33.3/41.6-MVA transformers. One transformer was constructed with a ribbon conductor winding and the other with a conventional winding. The ribbon conductor was installed in the high-voltage 141-kV winding. This design was a part of the EPRI advanced transformer project that began in 1987. The secondary winding was 13 kV and had a conventional conductor. According to ABB, core losses have been reduced by more than 10 percent and copper losses by 17 percent. Table 8-1 compares the weight and load and no-load losses of the ribbon conductor transformer against the conventional transformer. As can be seen from the table, advanced conductor design transformers not only have smaller losses than conventionally designed transformers but they weigh less as well. This is so because the advanced conductor takes up 30 percent less space.

Following the HL&P transformer, ABB manufactured and tested both an advanced conductor and a conventional designed 48/64/80-MVA transformer. The advanced conductor transformer had over 17 percent smaller no-load losses and over 10 percent smaller load losses than the convention conductor transformer (Table 8-2).

Several commercial advanced conductor designs have been manufactured since the HL&P prototype. They have exhibited calculated total loss savings of almost 10 percent. For example, a 45/60/75-MVA advanced conductor transformer had 9.4 percent smaller total losses than a conventional conductor transformer, while a 30-MVA FOA

Table 8-1. Comparison of 25/33/41.6-MVA Conventional and Advanced Transformers

Transformer characteristics	Conventional	Advanced	Improvement
Core and coil weight, lb (thousands)	197.0	186.2	5.5 percent
No-load loss, kW	21.5	19.3	10.4 percent
Load loss, kW	90.8	76.6	17.0 percent
Total loss	112.3	95.9	14.6 percent

SOURCE: R. S. Girgis, Design and performance improvements in power transformers using ribbon cable, *IEEE Transaction on Power Delivery*, vol. 10, no. 2, April 1995.

Table 8-2. Comparison of 48/64/80-MVA Conventional and Advanced Transformers

Transformer characteristics	Conventional	Advanced	Improvement
Core and coil weight, lb (thousands)	151.7	140.9	7.1 percent
No-load loss, kW	83.7	69.4	17.1 percent
Load loss, kW	82.0	73.4	10.5 percent
Total loss	165.7	142.8	13.8 percent

SOURCE: R. S. Girgis, Design and performance improvements in power transformers using ribbon cable, *IEEE Transaction on Power Delivery*, vol. 10, no. 2, April 1995.

advanced conductor transformer had 8 percent smaller losses. In addition to the advances in conductor design for reducing losses, there have been advances in the development of low-loss silicon steel.

Low-Loss Silicon Steel

What is a low-loss silicon steel core transformer? A low-loss silicon steel core transformer is a transformer with a core that is designed and manufactured to minimize no-load losses. Since utilities and some industrial users of transformers started evaluating losses in the 1970s, transformer manufacturers have been endeavoring to reduce the no-load losses of silicon steel core transformers. Figure 8-2 illustrates how the no-load losses of 50-kVA silicon steel core transformers have been going down since the early 1970s.

The transformer manufacturers have reduced the no-load losses of silicon steel core transformers by 50 percent in the last 30 years. Table 8-3 compares the no-load losses of low-loss silicon steel core transformers with in-service typical silicon steel core transformers. The manufacturers have accomplished this reduction in four ways. They

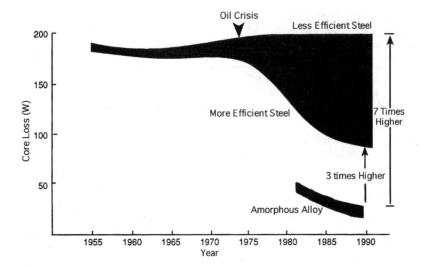

Many older silicon steel units on the electric utility systems have very high core losses when compared to both the moderately-high efficiency silicon steel transformers available today and the amorphous metal distribution transformers. An accelerated replacement program offers an opportunity to reap the benefits of increased energy savings.

Figure 8-2. Historical no-load loss trend for 50-kVA transformer. (*Courtesy of Allied Signal.*)

have been improving the construction of the silicon steel itself. They have been improving the cutting of the laminations. They have been improving the stacking or assembling of the laminations in the core of the transformer. And finally, by the use of improved computer models of the no-load losses, they have been able to design the core to reduce no-load losses.

Lamination Construction

One way to reduce no-load losses is to reduce the thickness of the core laminations. This reduces the eddy current losses by reducing the resistance between laminations. The reduction in thickness of grain-oriented silicon steel has been improving over the years.

The steady reduction in the thickness of grain-oriented silicon steel began with the introduction of the two-stage cold rolling process in 1934. Allegheny Ludlum Steel and Armco introduced even thinner-gauge steel in 1980. Since then, five main grades of silicon steel are being used in the transformer industry. They are identified as M2, M3, M4, M5, and M6 in increasing order of gauge from 7 to 14 mil thick-

Table 8-3. Comparison of Typical and
Low-Loss Silicon Steel-Core No-Load
Losses

	No-load losses (W)	
Rating (kVA)	Typical silicon	Low-loss silicon
Single-phase		
10	60	30
25	100	50
50	210	105
75	260	130
100	320	160
Three-phase		
75	370	185
150	540	270
300	950	475
500	1400	710
750	1750	875
1000	2400	1200
1500	3600	1800
2000	4000	2000
2500	4800	2400

SOURCE: Courtesy of GE.

ness. Figure 8-3 shows the no-load losses of each one of these gauges of silicon steel.

Laser scribing has been the latest development in reducing losses in silicon steel. In laser scribing, the laminations are sliced into strips parallel to the surface. This process orients the steel in the direction of the flux lines. This was developed by Nippon Steel and is more expensive than conventional cutting methods. Figure 8-4 shows how laser-scribed silicon steel losses compare with conventional silicon steel losses.

Eddy current losses are greatly reduced by constructing laminations that are clean and have burr-free edges. This removes the electrical connection between laminations. This is accomplished by more accurate machining of the laminations. In addition to improving the construction of the laminations themselves, improvement in stacking of the laminations has resulted in reduced eddy current losses.

By interleaving the laminations in groups of five to seven, eddy current losses are reduced. Rather than butting the laminations joints together, the interleaving of the laminations increases the amount of steel that bridges the joint gap. This results in reducing the resistance between the laminations and thus reducing the eddy current losses.

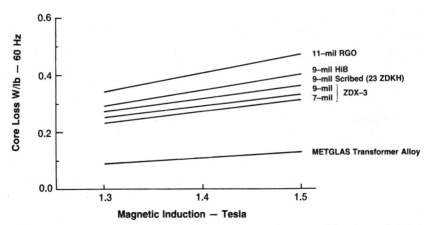

Figure 8-3. No-load losses for various silicon steel gauges. (*Courtesy of Allied Signal.*)

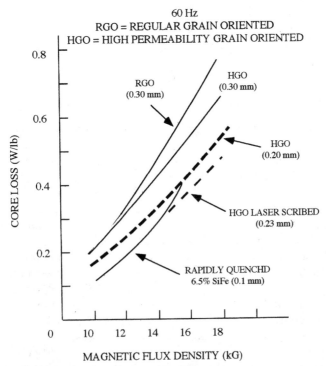

Figure 8-4. Laser-etched versus conventional silicon steel core no-load losses. (*Source: R. E. Luborsky and J. J. Becker, IEEE Transactions, MAG-15, 1979.*)

There have been improvements in the modeling of the magnetic field in transformers. This has resulted in a more accurate understanding of the core configuration and how it affects no-load losses. For example, in a 1993 IEEE paper entitled, "Calculation and Reduction of Stray and Eddy Losses in Core-Form Transformers Using a Highly Accurate Finite Element Modeling Technique," authors D. Pavik, D. C. Johnson, and R. S. Girgis reported that by using an accurate modeling technique, transformer manufacturers were able to determine what topologic changes could be made in the structural components and shielding members to minimize stray and eddy current losses. They said that "when end-frame aluminum shields were added, the `core edge' losses decreased by 29 percent. When magnetic tank shields were added, the core edge loss went down by an additional 55 percent."

The largest reduction in no-load losses came from the introduction of amorphous metal core distribution transformers in the 1980s. This has resulted in dramatic reductions in no-load losses by as much as 80 percent over older silicon steel core transformers. Chapter 9 provides a detailed analysis of how amorphous metal core transformers achieve these large reductions in no-load losses.

9

Amorphous Metal Distribution Transformers

Even though amorphous metal distribution transformers have been in operation since 1982, they are perceived as the latest advancement in the ongoing effort to improve the efficiency of transformers. Fourteen years of operational experience with amorphous metal transformers has demonstrated to both the transformer manufacturing and utility industries that amorphous metal transformers are interchangeable with traditional silicon steel core transformers. Although amorphous metal transformers significantly reduce losses, they may increase the life-cycle operating cost of the transformer due to the higher initial cost. Notwithstanding, if these new transformers are used to replace conventional transformer models, they could have a major impact on the operating cost of not only the United States but the world electric utility distribution systems. Replacement of the existing 40 million distribution transformers in the United States could reduce the estimated 52 billion kWh per year of distribution transformer core losses to 10 billion kWh, assuming that 80 percent of the core losses could be saved. In fact, in a recent brochure, the manufacturer of amorphous steel, Allied Signal, projected that by the year 2000, amorphous metal transformers could save the United States alone 47 billion kWh of annual energy. As shown in Table 9-1, Allied Signal estimates that transformer core losses for 12 major countries including the United States could increase to almost 160 billion kWh by the year 2000 and

Table 9-1. Potential Energy Savings of Amorphous Metal Distribution Transformers (AMDTs) Internationally

Country	Electricity consumption, 1994 (billion kWh)	Estimated distribution transformer core losses, 1994 (billion kWh)	Percent annual growth rate	Projected distribution transformer core losses, 2000 (billion kWh)	Potential savings with AMDTs, 2000 (billion kWh)
United States	3280	52	2	59	47
Europe	2960	31	1	33	26
Japan	930	15	2	17	14
China	840	15	8	24	19
Brazil	275	5	5	7	6
India	245	4	7	6	5
South Korea	147	3	10	5	4
Taiwan	106	2	8	3	2.5
Thailand	60	1	11	2	1.5
Indonesia	43	0.7	11	1.3	1.0
Malaysia	40	0.7	10	1.2	1.0
Philippines	26	0.5	14	1.1	0.9

Source: Courtesy of Allied Signal.

that amorphous metal transformers could save 80 percent or 125 billion kWh per year of these core losses. This amounts to $9.6 billion, assuming an energy rate of $0.075/kWh.

Can these savings be achieved cost-effectively? How do amorphous metal transformers attain these kinds of savings? What are amorphous metal transformers? *Amorphous metal transformers* refer to the makeup of the transformer core. *Amorphous* is defined by *Webster's Dictionary* as "unorganized, vague." This seems puzzling. How can cores of transformers be unorganized and vague? The definition goes on to say, "In chemistry and mineralogy, not crystalline." Now what do transformer cores have to do with crystals? This question can be answered best by looking at the atomic structure of amorphous metal compared with silicon steel.

The amorphous metal is not crystalline like silicon steel. This means that the atoms in amorphous metal are arranged in a different pattern than those in silicon steel. They are arranged in a noncrystalline random pattern, while silicon steel atoms are arranged in an orderly crystalline structure. This can be illustrated best by the diagram shown in Fig. 9-1 comparing the atomic structure of amorphous metal with that of silicon steel.

This noncrystalline molecular structure looks like glass and is often referred to as "glassy metal." Such "glassy metals" have unique magnetic properties. They exhibit very low hysteresis losses. This is so because when a metal is subjected to an alternating magnetic field, the atoms reorient themselves to be in alignment with the magnetic field and form tiny magnets. When the alternating field returns to zero, the atoms return to the same orientation as before the magnetic field was applied. As described in detail in Chap. 3, the movement of the atoms

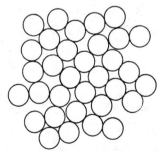

Crystalline silicon steel   **Noncrystalline amorphous metal**

Figure 9-1. Amorphous versus silicon steel atomic structure. (*Courtesy Allied Signal.*)

causes molecular friction that is given off in the form of heat and is called *hysteresis loss*. The randomness of the molecular structure of amorphous metal causes less friction than silicon steel when the magnetic field is applied to amorphous metal. The resulting reduction in hysteresis losses causes amorphous metal transformers to have significantly lower no-load losses than silicon steel core transformers. This ease of magnetization is illustrated in Fig. 9-2, which shows the magnetization curves of amorphous metal and silicon steel. As can be seen from these curves, amorphous metal saturates at a lower magnetization level than silicon steel. In addition to the reduction in hysteresis losses, amorphous metal transformers have lower eddy current losses.

Amorphous metal transformers have lower eddy current losses than silicon steel core transformers due to the very thin nature of the amorphous metal laminations. Silicon metal laminations are usually 7 to 12 mil in thickness, while amorphous metal laminations are less than 1 mil in thickness. Thin laminations result in low resistance between laminations and low eddy current losses.

The combination of reduced hysteresis and eddy current losses results in new transformers that use amorphous metal in their cores which lose 70 to 80 percent less energy in their core than silicon steel core transformers. For example, the core loss of a silicon steel core 25-kVA transformer is 57 W, while the core loss of an amorphous metal 25-kVA transformer is typically 15.4 W. Because these loss reductions are no-load losses, they occur 24 hours a day, 365 days a year. These loss reductions are offset by an increased cost of 25 to 30 percent for amorphous metal transformers as compared with silicon steel core

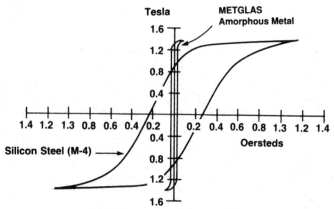

Figure 9-2. Magnetization curve for amorphous metals and silicon steels. (*Courtesy of Allied Signal.*)

transformers. How did these amorphous metal transformers come into being?

History

The historical development of amorphous metal distribution transformers is described in Fig. 9-3. In the late 1970s, as the importance of economic evaluation of core losses grew, a number of manufacturers began studying the suitability of amorphous metal alloys for use as a core material in transformers. Amorphous metal, first produced in the early 1960s, was thought to be an attractive core material for two reasons. First, it had excellent and unique magnetic properties, particularly very low electromagnetic losses that result from the glasslike, noncrystalline structure. This structure is a result of the cooling process. Second, the metal offers a potentially low production cost due to the simplicity of the manufacturing process: one-step continuous ribbon fabrication in contrast with the six or more steps needed for producing silicon steels.

Certain properties of the metal, however, increase the complexity of the manufacturing process. Amorphous metals are hard, brittle, and thin. Concern over the unique physical qualities of the metal led the Electric Power Research Institute (EPRI) to initiate a program in early 1983 to assess the commercial viability of using amorphous metal in transformer cores. The multistage project was designed to accelerate

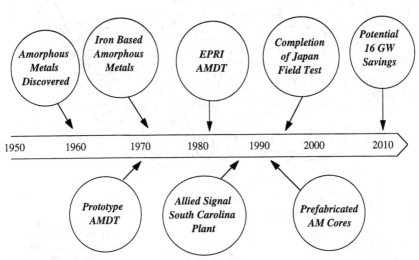

Figure 9-3. Historical development of amorphous metal distribution transformers (AMDTs). (*Courtesy of Allied Signal.*)

development of amorphous metal transformer cores. It first involved the manufacture and installation in 1982 of one 25-kVA distribution transformer on Duke Power's distribution system in Hickory, North Carolina, and the subsequent installation of 25 duplicate transformers throughout the country in 1983. It involved the evaluation of design alternatives, identification of fabrication and manufacturing requirements, a market analysis, and pilot-line production of an additional 1000 units for installation and testing. It was cosponsored with General Electric Company and the Empire State Electric Energy Research Corporation. Early work indicated that amorphous metal transformers could be expected to have a life cycle equal to that of conventional silicon steel core transformers.

In 1985, 1000 single-phase 25-kVA amorphous metal units were produced in the pilot manufacturing facility in North Carolina. They were then shipped to 90 participating EPRI member utilities throughout the country for a 2-year field trial (see map in Fig. 9-4). The purpose of this pilot project was to test manufacturing techniques and evaluate the field performance of amorphous metal transformers.

The field trial did confirm the long-term stability of low-loss amorphous metal transformers under actual operating conditions. Participants measured core losses and exciting currents on all units upon arrival and then on 10 percent of the units at 1-year intervals. All

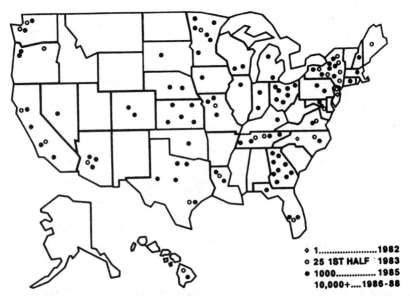

◇ 1........................1982
○ 25 1ST HALF 1983
● 1000...............1985
  10,000+....1986-88

Figure 9-4. Map of amorphous metal transformer pilot project participants and other GE amorphous metal transformer installations. (*Courtesy of GE.*)

units maintained their low no-load loss levels within ± 10 percent over the 2-year period. Table 9-2 is a summary of the no-load loss measurements of five participating organizations and illustrates how the no-load losses were maintained. The electrical and mechanical characteristics of these 1000 single-phase 25-kVA amorphous units are compared with those of silicon steel core transformers in Table 9-3.

The EPRI/GE field tests were a success. No failures occurred due to the amorphous metal core. Failures happened for other reasons. One unit failed from being overloaded 250 percent. This unit was returned to service, the winding was replaced, and the same amorphous metal core was reinstalled. Another unit was damaged when a car hit the utility pole it was mounted on. Other failures were caused by lightning and a leaky bushing.

Amorphous metal transformers have been available commercially since 1986. In April of 1989, Allied Signal began to operate the world's first commercial plant for the production of amorphous metal alloy in Conway, Arkansas. Several transformer manufacturers, including GE, ABB, and Howard Transformers, currently have been making amorphous metal core transformers. They include transformer sizes from 25-kVA single-phase to 2500-kVA three-phase units. The difference in no-load losses of these sizes of transformers for older silicon steel, low-loss silicon steel, and amorphous metal distribution transformers are shown in Table 9-4.

As can be seen from this table, the no-load losses of a silicon steel core transformer vary from 3 to 7 times higher than those of an amorphous metal transformer. The more recently developed silicon steel core transformers are highly efficient. With the advent of laser-etched silicon steel, the no-load losses of silicon steel core transformers have been reduced significantly. As can be seen from Fig. 8-2, the no-load loss difference between a 50-kVA silicon steel core transformer and an amorphous metal transformer from 1955 to 1990 decreased from 7 times greater to 3 times greater.

Allied Signal Inc. reported in a brochure by Patrick M. Curran and Dave Pendlebury titled "Amorphous Metal Distribution Transformers: A Cost Effective Technology for Environment Betterment" that "today [1995], over one half million transformers made with amorphous metal are in service on electric utility distribution systems around the world. The majority of these are in the United States, but the focus is changing to those regions with higher energy costs and more rapid growth in electricity demand." The Electric Power Research Institute and the State Department project that amorphous distribution transformers will have captured 25 percent of the market by the year 2000. This would result in an annual electric energy savings of 10 billion kWh and a 2.1 million metric ton reduction in carbon emissions, while utilities like Central Maine expect a net present-value savings of $335,000 in 10 years by installing 36,000 amorphous metal transformers.

Table 9-2. Summary of Amorphous Metal Transformer No-Load Loss Measurements

Organization	No. of transformers tested	Rating and type	Test period	Years in service	Average percentage change in core loss
Pacific G&E	9	1-phase 25-kVA poles	1986–87	1	+ 0.9
San Diego G&E	21	1-phase 25-kVA poles & pads	1986–87	1.5	–0.6
Allegheny PS	25	1-phase 10-kVA poles	1987–89	1.5	No change
Niagara Mohawk	1	3-phase 500-kVA	1986–89	2	–7.0
Naval Civil Engineering Labs	8	3-phase 75- & 150-kVA pads	1987–89	3	–0.25

SOURCE: Courtesy of GE.

Table 9-3. Comparison of Silicon Steel and
Amorphous Metal 25-kVA Transformers

Characteristic	Amorphous steel	Silicon steel
Core loss (W)	15.4	57
Load loss (W)	328	314
Exciting current (%)	0.14	0.36
Impedance (%)	2.45	2.45
Audible noise (db)	33	40
Temperature rise (°C)	48	57
Short-circuit test ($\times N$)	40	40
In-rush current ($\times N$)	12/13	23/14
(Calculation 0.01/0.1 s)		
TIF @100/110% exc.		
(IT/kVA)	2/10	5/25
Weight (lb)	441	406

SOURCE: W. D. Nagel, Ultra efficient amorphous transformer, *IEEE Transactions*, 1991.

Table 9-4. Comparison of Silicon Steel and Amorphous
Metal Transformer No-Load Losses

	No-load losses (W)		
Rating (kVA)	Typical silicon	Low-loss silicon	Amorphous
Single-phase			
10	60	30	11
25	100	50	20
50	210	105	32
75	260	130	39
100	320	160	54
Three-phase			
75	370	185	67
150	540	270	107
300	950	475	185
500	1400	710	260
750	1750	875	310
1000	2400	1200	420
1500	3600	1800	555
2000	4000	2000	750
2500	4800	2400	850

SOURCE: Courtesy of GE.

Approximately 1 million new distribution transformers are installed each year on U.S. utility systems. New low-loss amorphous metal distribution transformers offer a significant opportunity to increase system efficiency and defer generating capacity additions. If all new distribution transformers were amorphous metal transformers, this could result in energy savings of nearly 600 million kWh per year. If the estimated 40 million distribution transformers now in service in the United States were replaced with amorphous metal units, over 40 billion kWh of lost energy a year could be saved.

Present prices of amorphous metal transformers are 25 to 30 percent greater than those of comparable silicon steel core transformers. This price difference is expected to decrease as more amorphous metal is produced and new production techniques are developed. The effect of production on the price of amorphous metal transformers can best be understood by taking a closer look at the unique characteristics of amorphous metal production.

Manufacture

In the process of producing amorphous metal, a proprietary molten alloy (2300°F) of iron, boron, and silicon is cooled rapidly at a rate of about 1 million degrees Centigrade per second so that crystals cannot form. Two difficulties in the production of amorphous metal have had to be overcome. One is prevention of the metal from becoming crystalline. This has been overcome by developing a method for chilling the molten alloy at a rate that prevents it from becoming crystalline. The other barrier has been to develop a process that produces strips of amorphous metal that are wide enough to be used in transformer cores.

Several amorphous metal production processes have been developed over the last 20 years in an attempt to increase the metal strip width. The first process was the splat quenching process. This process was developed 30 years ago and produced millimeter-wide flakes with micrometer thicknesses. The next process was the droplet quenching process that produced only spherical flakes. The third process was the melt spinning process. In this process, the alloy is dropped on a chilled spinning wheel to allow the metal to cool quickly and prevent it from crystallizing. This produces a thin ribbon strip (0.025 mm) of metal. The continuous ribbon strip comes off the wheel at about 60 mi/h. The width of the strip, however, was still too narrow. Not until Allied Signal developed the planar flow casting process in the 1970s and overcame the width limitations could amorphous metal transformers be built. Table 9-5 summarizes the various processes developed for producing amorphous metal strips.

Table 9-5. Amorphous Metal Liquid Quenching Processes

Process	Examples	Product
Splat quenching	Gun Hammer and anvil	Flake
Droplet	Atomization Spark erosion	Powder
Surface melting	Laser Electron beam	Surface "glaze"
Melt spinning	Free jet spinning Twin roller Melt drag Planar flow casting	Continuous Wire and ribbon

SOURCE: Courtesy of Allied Signal.

The planar flow process is shown in Fig. 9-5. This process used the liquid alloy's surface tension to produce a strip of metal 6.8 in in width. This strip could be used to manufacture amorphous metal transformer cores.

Consequently, a pilot plant was built in 1986 by Allied Signal to produce the amorphous metal for the first trial transformers. Figure 9-6 is a schematic of that pilot plant. Based on this design, Allied Signal built a facility in Conway, Arkansas, that has the capacity of producing 60,000 tons of amorphous metal each year.

PLANAR FLOW CASTING

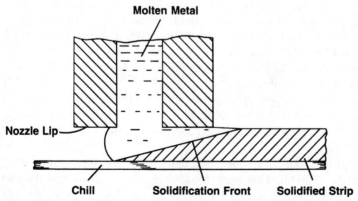

Figure 9-5. Planar flow casting process. (*Courtesy of Allied Signal.*)

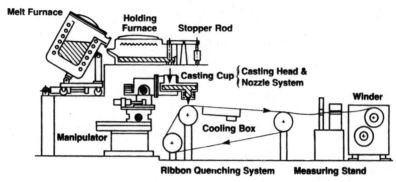

Figure 9-6. Schematic illustration of EPRI/Metaglas products pilot plant. (*Courtesy of EPRI.*)

New methods for producing amorphous metal continue to be developed. In a January 1993 *Electrical Construction & Maintenance Magazine* article entitled, "Choosing the Right Transformer," Robert B. Morgan reported on a new process that has promising results. In this article he said, "Recently, researchers at the University of Illinois discovered a procedure for making amorphous metals using a different method. When materials are treated with high-intensity ultrasound, tiny bubbles form and collapse rapidly in a process called *acoustic cavitation.* The Illinois researchers using this process, which also causes rapid heating and cooling, made amorphous iron that was 96 percent pure, compared to the conventional method of making the material with a purity of 80 percent. It is not known how practical the new process is from a commercial manufacturing standpoint."

After the amorphous metal has been produced, it is either assembled by Allied Signal into an amorphous metal transformer core or shipped to the transformer manufacturers to be assembled into a transformer core. The transformer manufacturers have to deal with two major problems of extreme hardness and thinness in assembling the amorphous metal into a transformer core.

They must deal with amorphous metal's extreme hardness. It is four to five times as hard as silicon steel. This characteristic causes cutting tools with carbide tips to wear out 1000 times faster than when used on silicon steel.

The design of amorphous metal transformers involves minimizing the cutting of the metal. This leads to winding the coils around the coil continuously. Figure 9-7 shows the various core design concepts that were considered in the early development of the amorphous metal

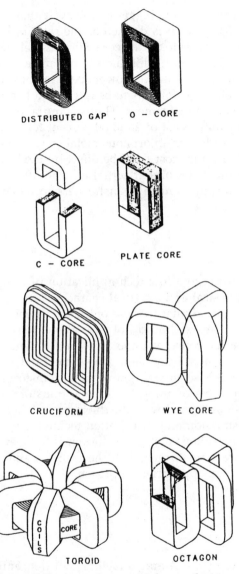

DISTRIBUTED GAP   O - CORE

C - CORE   PLATE CORE

CRUCIFORM   WYE CORE

TOROID   OCTAGON

Figure 9-7. Amorphous metal core designs.
(*Courtesy of GE.*)

core. GE decided to use the distributed gap construction for the 1000
25-kVA units constructed for the EPRI pilot project. Allied Signal's core
design has a toroidal, continuously wound core.

The other problem is the lower space factor of the thin, almost foil-
like laminations of amorphous metal. The low space factor results in

the amorphous metal transformers being bigger and heavier than silicon steel core transformers. *Space factor* is defined as the ratio of core cross-sectional area to the area available for the core. Silicon steel laminations have a space factor of 96 percent, while amorphous metal laminations have a space factor of 80 percent. This lower space factor causes the cores of amorphous metal transformers to be larger and heavier than those of silicon steel core transformers. These manufacturing problems result in the higher initial cost of 25 to 30 percent for amorphous metal transformers. In order for amorphous metal transformers to be cost-effective, the higher initial cost must be offset by the lower no-load losses. How much must the higher initial cost be offset by lower no-load losses for amorphous metal transformers to be cost-effective?

Economics

Amorphous metal transformers are a low-tech application of a high technology. As long as the physical and electrical characteristics of an amorphous metal transformer are similar to those of a silicon steel core transformer, the transformer users should not care about the core material. They should instead be concerned about the performance of the amorphous metal transformer. They should be concerned with whether it is economical to purchase an amorphous metal transformer. In evaluating the cost-effectiveness of amorphous metal transformers, the transformer purchaser is confronted with the dilemma of whether to treat them as a new transmission and distribution technology or a conservation option. If the amorphous metal transformer is to be evaluated like a silicon steel core transformer, then the utility standard method of total owning cost is an appropriate approach.

Total Owning Cost Economics

Utilities know that the value of the A and B factors will determine whether amorphous metal transformers are cost-effective. If the A factor for evaluating no-load losses in particular is large enough, the higher initial cost of amorphous metal transformers will be justified. This can best be illustrated by an example of a typical 25-kVA amorphous metal distribution transformer compared with a low-loss silicon steel core distribution transformer.

An example from an 1987 EPRI journal article entitled, "Transformers with Lower Losses," shows the break-even point for the A and B

Table 9-6. Comparison of Silicon Steel and Amorphous Metal Transformer Total Owning Costs

TOC components	25-kVA cost	
	Low-loss silicon steel	Amorphous metal
No-load losses	48 W	18 W
	×$5/W	×$5/W
	$240	$90
Load losses	284 W	249 W
	×$1/W	×$1/W
	$284	$249
Total losses	$525	$339
Purchase price	$510	$695
Total owning cost	$1034	$1034

SOURCE: Harry Ng and Bill Shula, Transformers with lower losses, *EPRI Journal,* October–November 1987.

factors in the total owning cost calculation assuming that the cost of a 25-kVA amorphous metal transformer is $695 and that the cost of a low-loss silicon steel core transformer is $510. For the utility that values no-load losses at $5 per watt and load losses at $1 per watt (present worth over the life of the transformer), the break-even point in total owning cost for both types of transformers is shown in Table 9-6.

Conservation Economics

Amorphous metal transformers require very little technical understanding of how they work to use them. They also can be treated as a demand-side management conservation option. If they are owned by the utility, they have the advantage over other demand-side management conservation options of not reducing the utility's revenue. If they are owned by the end-user, they can be treated as a conservation option, as in replacing incandescent light bulbs with fluorescent lights or replacing an existing water heater with a more efficient unit.

Amorphous metal transformers as a conservation option have several advantages over conventional alternatives. They have the advantage that the savings can be determined easily from test results. The no-load loss savings are not dependent on the loading and use of the transformer. The savings occur over the 30-year life of the transformer. The savings are not dependent on the behavior of the utility or end-user. The savings do not result in any revenue loss to the utility. In fact, the savings result in a revenue gain to the utility. Finally, use of the amorphous metal transformer does not affect the comfort of the end-user, as in reducing the cooling setting of air conditioning or reducing the heating setting of an electric furnace.

There are basically three ways to evaluate the cost-effectiveness of conservation options. They are the *total resource cost test*, the *utility cost test*, and the *ratepayer impact measure test*. The total resource cost method measures the net cost of the option, including the end-user and utility costs. The utility cost method measures the net cost of the conservation alternative based on costs incurred by the utility, including incentive payments but excluding participant costs. The ratepayer impact measure method measures the impact on customer bills to changes in utility operating costs and revenues caused by the conservation option. These methods were developed by the California Public Utility Commission and are contained in the December 1987 *Standard Practice Manual for Economic Analysis of Demand-Side Management Programs.*

Each one of these economic methods was used by EPRI in the EPRI DSManager computer program to compare amorphous metal transformers with traditional demand-side management options. This comparison is contained in an October 1994 EPRI report (TR-104246) entitled, *Cost-Effectiveness Analysis of Amorphous Core Transformers Using EPRI DSManager.* For example, in this study, EPRI used the ratepayer test to compare the benefit-cost ratio of energy savings associated with 25-, 75-, 500-, and 2000-kVA amorphous metal transformers (AMTs) with efficient water heaters, compact fluorescent lights, direct load control air conditioning, fluorescent lighting, efficient motors, and standby generation. As can be seen from Fig. 9-8, the benefit-cost ratio

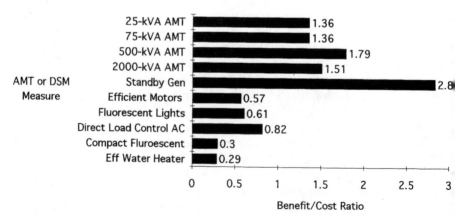

Figure 9-8. Amorphous metal transformers (AMTs) versus DSM options. (*Courtesy of EPRI.*)

of the amorphous metal transformers varies from 1.35 to 1.51. The transformers are assumed on the utility side of the meter, while all the comparable conservation options are less than unity except for the 2.84 benefit-cost of standby generation and are assumed on the end-user side of the meter.

The Bonneville Power Administration (BPA) sponsored a study of various technologies for improving the efficiency of transmission and distribution systems. BPA is a federal power marketing agency within the Department of Energy. It sells power to over 150 utilities in the Pacific Northwest and California primarily from the output of hydro-electric dams on the Columbia River. The purpose of the study, i.e., the Transmission and Distribution Efficiency Improvement R&D Survey Project, was to identify and quantify technologies that would provide the greatest benefits in improving utility system efficiency. The technologies were subjected to an evaluative test in the late 1980s using a modeled sample system for Benton County PUD. Benton County PUD is a utility in Washington State that has characteristics typical of a BPA utility customer's system. Of the 15 state-of-the-art technologies modeled, the most significant reduction in losses was achieved through the replacement of standard transformers with amorphous metal units. Over 35 percent of the peak kilowatt losses, or about 719 kW, were eliminated. BPA also participated in the 1000-unit amorphous metal 25-kVA transformer pilot test program sponsored by EPRI. BPA installed 25 units in several substations and experienced no change in the no-load losses over the 2-year test period. Based on these studies, BPA adopted a policy of requiring all its purchases of distribution transformers to be amorphous metal transformers.

BPA is a bulk power transmitter of electrical energy, and consequently, its primary use of distribution transformers is for station service (i.e., 300 existing station service distribution transformers in operation). As with other utilities, BPA had three opportunities for the application of amorphous metal transformers. The first opportunity presents itself when existing transformers are replaced due to age or damage. A second opportunity exists in PCB-filled (insulated) transformers. Amorphous metal transformers have been used to replace any existing distribution transformers that have been found to exceed the 50 parts per million PCB contaminant limit. Last, as technical and economic performance is demonstrated for amorphous metal transformers with larger kilovoltampere ratings, opportunities for new and replacement of existing transformers will be considered.

BPA also introduced an amorphous metal transformer rebate program to encourage its utility customers to include amorphous metal transformers as another conservation option. The rebate program was adopted at the beginning of 1994. The rebates provided in Table 9-7 are based

Table 9-7. Single-Phase Amorphous Metal
Transformer Rebate Schedule in 1990 Dollars

Transformer rating (kVA)	Maximum load losses (W)	Maximum no-load losses (W)	Rebate ($)
10	170	15	70
15	210	31	75
25	280	25	125
37.5	340	35	150
50	510	40	175
75	740	45	200
100	960	60	275
167	1130	90	350

SOURCE: Courtesy of Bonneville Power Administration.

on a calculation of one-half the estimated incremental cost of low-loss silicon steel core transformers and the cost of amorphous metal distribution transformers. The data used to estimate costs came from a variety of sources, including utility records and vendor quotes. A minimum load loss efficiency was set to prevent the possible increased load losses of amorphous metal transformers from offsetting the saving of no-load losses. The rebate values were determined by the following formula:

$$\text{Rebate} = \text{estimated price differential} - 1/2 \text{ present value of BPA's rate for 30 years} \times \text{annual kWh savings} \tag{9.1}$$

$$\text{Maximum rebate} = \text{present value of a 30-year resource at } 42 \text{ mills/kWh} \times \text{annual kWh savings} \tag{9.2}$$

BPA did not include any single-phase transformers larger than 167 kVA or any three-phase transformers. The cost and energy savings of these transformers caused them to exceed the maximum rebate of 42 mills/kWh.

BPA has since stopped all its conservation incentive programs, including the amorphous metal transformer rebate program. It now encourages its utility customers to adopt their own conservation programs.

Conclusions

The amorphous metal distribution transformer has proven to be a reliable and energy-efficient technology. In a November 1991 IEEE article entitled, "Amorphous Alloy Core Distribution Transformers," the authors Harry Ng, Ryusuke Hasegawa, Albert Lee, and Larry Lowdermilk estimate that there have been 60,000 to 70,000 units purchased and installed throughout the world. Amorphous metal transformers continue to make inroads into the transformer market. There have been some technical concerns besides its extra cost of 25 to 30 percent over a conventional silicon steel core transformer that have affected its marketability. These technical problems include extra weight and ferroresonance.

Weight

The lower space factor of amorphous metal cause the amorphous metal core to take up more space than a silicon steel core. The lower saturation level of amorphous metal also requires more metal than a conventional silicon steel core. This results in amorphous metal transformers tending to be heavier and larger than silicon steel core transformers. Table 9-8, from a November 1991 IEEE article entitled, "Amorphous Alloy Core Distribution Transformers," illustrates the increased weight of amorphous metal transformers over silicon metal core transformers.

Table 9-8. Comparison of Silicon Steel and Amorphous Metal Core Transformer Weights

	Weight (pounds)	
Rating (kVA)	Amorphous metal	Silicon steel
Single-phase		
10	318	300
15	422	321
25	441	406
50	719	709
75	994	821
100	1131	961
Three-phase		
75	2030	2000
150	2870	2900
300	4360	3600
500	6090	4900
750	6600	6800

SOURCE: Harry Hg, Ryusuke Hasegawa, Albert Lee, and Larry Lowdermilk, Amorphous alloy core distribution transformers, *IEEE Transactions*, 1987.

The increased size and weight can cause utilities difficulty in installing amorphous metal transformers on existing poles and structures designed for lighter and smaller units. One way of handling this extra cost is to add the cost of larger pads to the transformer evaluation. Another way is to specify weight and size limitations. For example, one large purchaser of amorphous steel core transformers, Central Maine Power Company, specifies that its 25-kVA distribution transformers cannot weigh more than 500 lb. This encourages the transformer manufacturers to design amorphous metal transformers to be lighter and smaller. Another technical problem with amorphous metal transformers and low-loss silicon steel core transformers is their tendency to go into ferroresonance.

Ferroresonance

Ferroresonance in transformers occurs when the inductive reactance of the transformer resonates with the capacitive reactance of the feeder. This is especially true when the feeder is long and underground. The longer underground feeder increases the capacitive reactance of the feeder. The resonant condition can result in ferroresonant overvoltages. In the past, the high no-load losses in older transformers would attenuate the ferroresonant overvoltages. Now, with modern low-loss silicon steel and amorphous metal transformers, ferroresonant overvoltages are not attenuated. A low-loss transformer is susceptible to ferroresonance when any switching occurs near the transformer. The solution to this problem is the application of an outside arrester mounted on the transformer or a liquid-immersed arrester mounted inside the transformer. The arrester clamps the ferroresonant overvoltage and drops the transformer out of ferroresonance. The cost of an internal arrester can add $100 to the cost of the distribution transformer.

Amorphous Metal Substation Transformers

The Electric Power Research Institute and Westinghouse attempted in the 1980s to develop an amorphous metal substation transformer. This project has since been terminated. Techniques for assembling large laminations of amorphous metal for use in substation transformers have yet to be developed. Until these techniques have been developed, amorphous metal transformers will continue to be limited to distribution transformer sizes.

10
K-Factor
Transformers

K-factor transformers came into existence about 5 years ago out of necessity. They were introduced as a solution to the problem of harmonics on transformers. They are not necessarily more efficient than non-*K*-factor transformers. Harmonics on transformers cause increased losses and heating. Heating deteriorates the insulation and may cause the eventual failure of the transformer. With the increased use of microprocessors and nonlinear devices, such as adjustable-speed drives, harmonics have become an increasing phenomenon. These harmonics can be either eliminated by the use of filters or carried by specially designed *K*-factor transformers.

The *K*-factor value can be used to pick a specially designed *K*-factor transformer or to derate a non-*K*-factor transformer. The steps in calculating the *K*-factor value will be presented in this chapter along with examples. Then the transformer user can decide whether to derate a standard transformer or purchase a specially designed *K*-factor transformer. If a standard transformer is purchased, the *K* factor is used to derate the transformer. If a *K*-factor transformer is purchased, the *K* factor determines the *K*-factor rating of the transformer. The *K*-factor transformer is designed to carry the additional losses and heating caused by harmonics. United Laboratories (UL) Standards 1561, *Dry Type General Purpose and Power Transformers*, and 1562, *Transformers, Distribution, Dry-Type-Over 600 Volts*, have restricted the use of non-*K*-factor transformers to loadings of less than 5 percent harmonic content. UL provides a standard for *K*-factor ratings of 1, 4, 9, 13, 20, 30, 40, and 50. What are harmonics, and how do they cause losses and overheating in transformers?

Harmonics

Harmonics are multiples of the 60-Hz fundamental voltage and current. They add to the fundamental 60-Hz waveform and distort it. They can be 2, 3, 4, 5, 6, 7, etc., times the fundamental. For example, the third harmonic is 60 Hz times 3, or 180 Hz, and the sixth harmonic is 60 Hz times 6, or 360 Hz. The waveform in Fig. 10-1 shows how harmonics distort the sine wave.

Harmonics usually come from nonlinear devices. A *nonlinear* device is a device that draws a nonsinusoidal current when a sinusoidal voltage is applied. Nonlinear devices include arcing devices such as arc furnaces, saturable devices such as transformers, and power electronics equipment such as adjustable-speed drives. Yes, even transformers generate third harmonics. Figure 10-2 illustrates the various nonlinear loads and the corresponding harmonic waveforms they generate.

Harmonics' Effect on Transformers

Harmonics' major effect on transformers is to increase losses and heating. They increase both load and no-load losses. They increase load

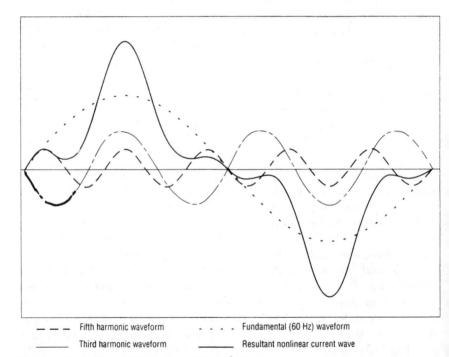

— — — Fifth harmonic waveform	· · · · Fundamental (60 Hz) waveform	
———— Third harmonic waveform	———— Resultant nonlinear current wave	

Figure 10-1. Composite harmonic waveform.

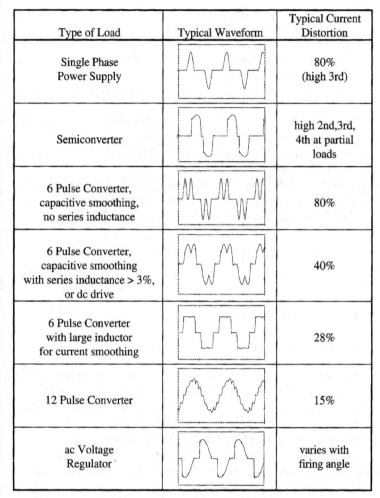

Type of Load	Typical Waveform	Typical Current Distortion
Single Phase Power Supply		80% (high 3rd)
Semiconverter		high 2nd,3rd, 4th at partial loads
6 Pulse Converter, capacitive smoothing, no series inductance		80%
6 Pulse Converter, capacitive smoothing with series inductance > 3%, or dc drive		40%
6 Pulse Converter with large inductor for current smoothing		28%
12 Pulse Converter		15%
ac Voltage Regulator		varies with firing angle

Figure 10-2. Nonlinear loads and their current waveforms. (*Courtesy of EPRI.*)

losses by causing skin effects, increasing eddy current, I^2R, and stray losses. They increase no-load losses by increasing hysteresis losses.

Load Losses.　　Load losses include I^2R, winding eddy current, and stray losses. Harmonics' effect on load losses and how to calculate the increased load losses are described in ANSI/IEEE C57.110-1986, entitled, *IEEE Recommended Practice for Establishing Transformer Capability When Supplying Nonsinusoidal Load Currents.* In this standard, the following formula is presented:

$$P_{LL} = I^2R + P_{EC} + P_{OSL} \qquad (10.1)$$

where P_{LL} = total load losses
I^2R = current squared times conductor resistance
P_{EC} = winding eddy current losses
P_{OSL} = other stray losses in clamps, core, and structural elements

As can be seen from this formula, the total load losses are equal to the I^2R losses plus the winding eddy current losses plus stray losses. Examining each component of this total loss formula helps explain how harmonics affect the total load losses.

Harmonics increase I^2R losses by increasing the total current I. This is so because harmonic currents add to the fundamental 60-Hz current. The rms value of the load current can be determined from the following formula:

$$I = \left[\sum_{h=1}^{h=h_{max}} (I_h)^2 \right]^{1/2} \qquad (10.2)$$

where I = rms value of load current
h = harmonic order (1, 2, 3, 4, etc.)
I_h = rms load current at the harmonic order h

Harmonics increase eddy current losses. Eddy current losses are proportional to the square of the load current and the square of the frequency of the harmonic. These losses can cause excessive winding losses and heating if significant harmonic load currents are present. Eddy current losses are caused by eddy currents that are produced by stray magnetic flux outside the transformer core. Eddy currents flow perpendicular to the load currents. The increased flux density caused by harmonics causes more stray magnetic flux and eddy current losses. The following formula describes the relationship between eddy current losses and harmonics:

$$P_{EC} = P_{EC-R} \left[\sum_{h=1}^{h=h_{max}} (I_h h)^2 \right] \qquad (10.3)$$

where P_{EC} = eddy current losses due to any defined harmonic load current
$P_{EC}-R$ = eddy current losses under rated conditions for a transformer winding
h = harmonic order (1, 2, 3, 4, etc.)
I_h = rms load current at the harmonic order h

Harmonics also affect other stray losses (P_{OSL}) in the core, clamps, and structural parts, although this component is small and will be neglected.

If the stray losses are neglected, then the per unit total load current P_{LL} under harmonic current conditions can be expressed by the following equation:

$$P_{LL} = \sum I_h^2 + P_{EC-R} \left[\sum_{h=1}^{h=h_{max}} (I_h h)^2 \right] \qquad (10.4)$$

where P_{EC-R} = eddy current loss factor under rated conditions for a transformer winding

h = harmonic order (1, 2, 3, 4, etc.)

I_h = per unit load current at the harmonic order h

In addition to the harmonics effect on load losses described in ANSI/IEEE C57.110-1986, there are increased losses caused by skin effect. Normally, when a transformer conductor is carrying 60-Hz current, the current is confined to the interior of the conductor. This is not the case with the increased frequency of harmonics. Instead, the increased frequency of harmonics results in the current flowing on the surface of the conductor. This phenomenon is called *skin effect* and causes increased losses and overheating. This is so because the resistance on the surface of the conductor is greater than the resistance inside the conductor. Therefore, as more current flows on the surface of the conductor due to skin effect, conductor resistance losses and heating increase. Losses due to skin effect often are caused by a third harmonic (180 Hz) flowing in the neutral conductor.

No-Load Losses.   Harmonics affect no-load losses by increasing hysteresis losses. They cause this by increasing the flux density and the rate of change in the core magnetization and demagnetization. This is so because the flux density and the rate of change are directly proportional to the frequency of the applied voltage. As the frequency increases, the flux density and the rate of change increase. The increased flux density increases the resistance to magnetization and the hysteresis losses. The increased rate of change to the core causes more energy to be required to magnetize the core and also increases hysteresis losses.

K Factor

K factor is a constant developed to take into account the effect of harmonics on transformer loading and losses. *K* factor is not mentioned in ANSI/IEEE C57.110-1986. However, ANSI/IEEE C57.110-1986 provides the methods for calculating the losses and currents for a certain harmonic load that is the basis for determining K-factor value. This standard was intended for application with liquid-immersed transformers, yet K factor as derived from this standard is applied to dry-type transformers. This implies some inaccuracy in the concept of K factor used with dry-type transformers.

The purpose of the K-factor rating is to rank transformers for harmonics, reduce skin-effect losses, and reduce the possibility of core saturation. Transformers with a K-factor rating have a note on their nameplates indicating that they are designed for nonsinusoidal current load with a certain K factor. K factor is defined as the ratio of eddy current losses divided by the harmonic current squared and is represented by the following formula:

$$K = \frac{\sum (I_h h)^2}{\sum (h)^2} \tag{10.5}$$

where I_h = harmonic current
      h = harmonic value

As can be seen from this formula, the K factor is determined by summing the product of each harmonic current squared and the harmonic order squared and dividing by the summation of the harmonic order squared. This summation is multiplied by the rated eddy current losses to obtain the increased eddy current losses due to harmonics.

The steps in calculating the K factor of a transformer are contained in the flowchart in Fig. 10-3. This flowchart provides a step-by-step method for determining the K factor to be used either to derate a standard transformer or to specify the K rating of a K-factor transformer. There are basically four steps for derating a transformer:

Step 1: In this step, it is necessary to determine the order of the harmonics and corresponding percentage harmonic content of the total load current. This can be determined from actual harmonic measurements or by the use of signature waveforms for various harmonic loads. Waveforms and a table of their harmonic content are provided in Figs. 10-4 to 10-8 for typical harmonic loads. These harmonic loads include the switched-mode power supply load, fluorescent lighting load, ac adjustable-speed drive without an input choke (filter), ac adjustable-speed drive with a 3 percent choke, and a dc drive. The transformer impedance can affect the waveform and harmonic content of the load current. This is so because the transformer impedance acts as an inductance choke to reduce the harmonic content.

Step 2: Calculate the K factor using Eq. 10.5 for each harmonic by summing the product of the square of the harmonic order and the har-

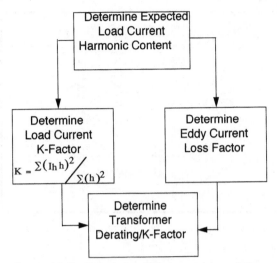

Figure 10-3. Calculating *K* factor and transformer derating flow diagram.

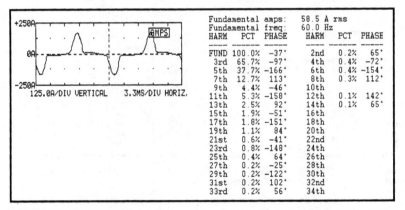

Figure 10-4. Current waveform and spectrum for switched-mode power supply load. (*Courtesy of EPRI.*)

monic current squared and dividing by the sum of the harmonic current squared. The worksheet contained in Table 10-1 on page 156 facilitates the calculation of the *K* factor for various harmonic conditions (the blank values need to be filled in for each case). This table can be constructed easily by using a spreadsheet computer program. The current, I_h, is in per-unit, or PU, of load current at harmonic order *h*.

Step 3: Determine the transformer eddy current loss factor. The transformer eddy current loss factor is a measure of the transformer

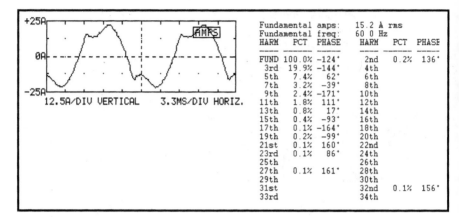

Figure 10-5. Current waveform and spectrum for fluorescent lighting load. (*Courtesy of EPRI.*)

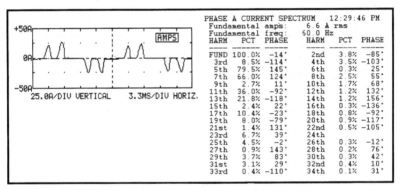

Figure 10-6. Current waveform and spectrum for ac adjustable-speed drive without an input choke load. (*Courtesy of EPRI.*)

eddy current losses. It can be obtained from the transformer designer, transformer test data, and the method in C57.110, or from typical values based on transformer type and size. Table 10-2 contains typical values obtained from D. E. Rice's 1984 IEEE paper entitled, "Adjustable-Speed Drive and Power Rectifier Harmonics: Their Effects on Power System Components."

Step 4: Determine the transformer derating by calculating the maximum current using the eddy current loss factor calculated in step 3 and the *K* factor for load current calculated in step 2. Determine the maximum current according to the following formula

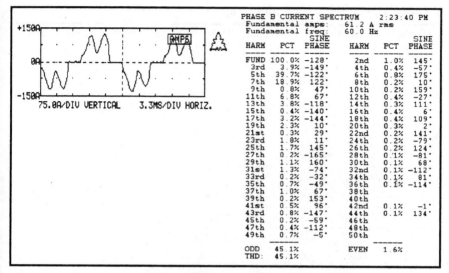

Figure 10-7. Current waveform and spectrum for ac adjustable-speed drive with a 3 percent choke inductance load. (*Courtesy of EPRI.*)

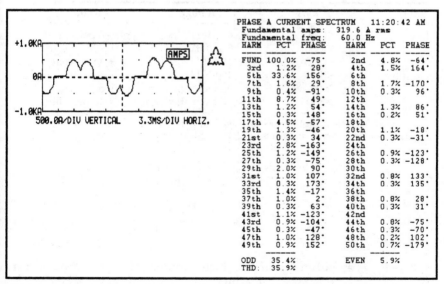

Figure 10-8. Current waveform and spectrum for a dc drive load. (*Courtesy of EPRI.*)

Table 10-1. K Factor Calculation

Harmonic number (h)	Frequency f (Hz)	Current I (pu)	I^2 (pu)	$I^2 \times h^2$
1	60			
3	180			
5	300			
7	420			
9	540			
11	660			
13	780			
15	900			
17	1020			
19	1140			
21	1260			
23	1380			
25	1500			

Table 10-2. Typical Values for P_{EC-R} to Use for Transformer Derating Calculations

Type	kVA	Voltage	Percent P_{EC-R}
Dry	≤1000	480-V HV	3–8 percent
	≥1500	5-kV HV	12–20 percent
	≥1500	15-kV HV	9–15 percent
Oil-filled	≤1000	480-V LV	1 percent
	2500–5000	Not specified	1–5 percent
	>5000	Not specified	9–15 percent

SOURCE: D. E. Rice, Adjustable-speed drive and power rectifier harmonics: Their effects on power system components, *IEEE* PCIC-84-52.

$$I_{\max} = \sqrt{\sum I_h^2} = \sqrt{\frac{1 + P_{EC-R}}{1 + K \times P_{EC-R}}} \qquad (10.6)$$

where P_{EC-R} = eddy current loss factor under rated conditions for a transformer winding

h = harmonic order (1, 2, 3, 4, etc.)

I_h = per unit load current at the harmonic order h

$$K = \frac{\sum (I_h h)^2}{\sum (h)^2}$$

An example of a large personal computer (PC) load should help clarify the previous procedure for calculating the derating of a transformer due to harmonics.

Example A commercial customer has a floor of a building with 200 PCs. The total load is approximately 100 kW. It is served from a single 150-kVA transformer. What is the K factor and the transformer derating?

solution The first step in finding a solution to this example is to determine the harmonic content for PC loads from Fig. 10-4. The next step is to insert these numbers into Table 10-1 to calculate the K factor, as shown in Table 10-3. Finally, the maximum current is calculated after determining from Table 10-2 that the eddy current loss factor (P_{EC-R}) is 8 percent:

$$I_{max} = \sqrt{\sum I_h^2} = \sqrt{\frac{1 + 0.08}{1 + (6.3 \times 0.08)}} = 0.85 \qquad (10.6)$$

Therefore, the transformer must be derated 85 percent, or 127.5 kVA. Or better yet, a K-factor transformer is used with a K rating of 7.

Table 10-3. K-Factor Example

Harmonic, h	Current (%)	Frequency (Hz)	Current (pu)	I^2	$I^2 \times h^2$
1	100.00	60	1	1	1
3	65.70	180	0.657	0.431649	3.884841
5	37.70	300	0.377	0.142129	3.553225
7	12.70	420	0.127	0.016129	0.790321
9	4.40	540	0.044	0.001936	0.156816
11	5.30	660	0.053	0.002809	0.339889
13	2.50	780	0.025	0.000625	0.105625
15	1.90	900	0.019	0.000361	0.081225
17	1.80	1020	0.018	0.000324	0.093636
19	1.10	1140	0.011	0.000121	0.043681
21	0.60	1260	0.006	0.000036	0.015876
23	0.80	1380	0.008	0.000064	0.033856
25	0.40	1500	0.004	0.000016	0.01
Total				1.596199	10.108991
K factor				6.333165	

A K-factor transformer is a specially designed transformer that has certain features that allow it to handle the extra heating of harmonic currents. It may have a static shield between the high- and low-voltage windings to reduce electrostatic noise caused by harmonics. It may use smaller than normal transposed and individually insulated conductors to reduce the skin effect and eddy current losses. It also may have a larger neutral conductor in the secondary winding to handle the third harmonic neutral currents. The core laminations may be individually insulated to reduce eddy currents in the core. A larger core with special steel may be required to reduce hysteresis losses and reduce the possibility of the transformer saturating due to high peaks on the distorted bus voltage waveform. The special steel has less resistance to the changing magnetic fields. A larger core increases the area of steel and thus reduces the flux density and resistance to the changing magnetic fields. Another way of reducing the flux density is to increase the number of turns in the winding. Heating due to increased I^2R losses is reduced by the use of larger conductors to reduce the conductor resistance. Often, cooling ducts are added to the windings to reduce the increased heating effects of harmonics. Eddy current losses are reduced by reducing the height of the conductor and decreasing the flux density, as discussed previously.

A K-factor transformer may cost approximately twice as much as a standard transformer and weigh 115 percent more than a standard transformer. It is still recommended that it be used rather than derating a standard transformer. This is so because a derated transformer may still contain hot spots due to harmonics that could result in overheating and transformer loss of life.

There are some recommended practical considerations in determining the K factor of a particular load. They include the following:

1. K-factor calculations are based on system measurements. When measurements are not available, determining a realistic K factor can be difficult.

2. Single-phase harmonic sources may inject harmonic currents into the system that can be out of phase with other currents at that same frequency. This results in some cancellation of the total harmonic current at that frequency.

3. Due to harmonic cancellation, although the K factor of an individual load may approach 30, the K factor for a circuit devoted to several of these types of loads may rarely exceed 9.

4. Transformer losses will increase with the frequency of the current; therefore, high-frequency harmonic components can be more important than lower-frequency components in causing transformer heating.

In conclusion, it is recommended that consideration be given to the following requirements when specifying K-factor transformers:

1. Transformer K-factor ratings must be matched to the K factor of the load and circuit being served.

2. Transformer manufacturers do not build transformers for every possible *K*-factor rating. Manufacturers normally will standardize their *K*-factor-rated transformers according to a finite set of *K*-factor ratings.

3. The transformer *K*-factor rating must either be equal to or greater than the *K* factor of the load and circuit the transformer serves.

4. *K*-factor-rated transformers will have lower impedance than non-*K*-factor-rated transformers.

5. For most application of *K*-factor transformers, a *K*-factor rating of 13 or less is adequate.

6. Transformer *K*-factor ratings are based on UL Standards 1562 and 1562.

11
Future Trends

What are the future trends of energy-efficient transformers? Will they become even more efficient and less costly? What new technologies are likely to be developed in the future? Will new technologies make them more cost-effective? Will there be increased or reduced use of energy-efficient transformers? Will the future use of energy-efficient transformers be driven by improved technology, economic incentives, regulatory mandates, or all these factors?

How can utilities or commercial and industrial users of transformers make a decision that takes into account future trends? If they incorrectly anticipate the future, they may be stuck with a 30-year mistake. If they correctly anticipate future trends, they could save money and energy that would make them more competitive with other utilities.

How will future trends affect the transformer manufacturing industry? Will the industry continue to develop new, energy-efficient transformer technologies? Will the industry reduce or increase its energy-efficient transformer research and development efforts? Will it continue to develop ways to reduce the cost of manufacturing energy-efficient transformers? How will it respond to the changes occurring in the electric utility industry?

How will research organizations such as the Electric Power Research Institute (EPRI) respond to the changes in the utility industry? Will it continue to fund research and development that improves the efficiency of transformers? How will its research and development priorities be affected by the changes in the electric utility industry?

This chapter will attempt to answer these and other questions about the future trends in the use of energy-efficient transformers. Future trends are driven by the three factors of competition, technology, and deregulation. The purpose of this chapter is to examine how the future trends of competition, technology changes, and deregulation will affect the use and makeup of energy-efficient transformers in the foreign and domestic markets. One of today's major uncertainties and tomorrow's future trends that could have the greatest effect on the use of energy-efficient transformers is the deregulation of the electric utility industry.

Deregulation

Deregulation of the electric utility industry in the United States seems to be an inevitable future trend that will become a reality in the next century. Deregulation has been in effect for several years in many parts of the world, including the United Kingdom, Australia, New Zealand, and South America. Deregulation in the United States is a relatively new phenomenon. In fact, many U.S. utilities are purchasing deregulated foreign utilities in order to get experience in how to compete in the upcoming deregulated utility market. The deregulation process began in the United States with the passage of the 1992 Energy Policy Act. Passage of this act was soon followed by the Federal Energy Regulatory Commission (FERC) introducing on April 7, 1995, the Notice of Proposed Rulemaking (Mega-NOPR).

With the Mega-NOPR, FERC requires utilities to provide open transmission access and separate their power business from their transmission and distribution (t&d) business. This has an effect on the electric utility industry in the United States. In the past, the electrical utility industry was dominated by so-called vertically integrated monopolies. This means that utilities owned generation, transmission, and distribution facilities and provided electrical energy to designated franchised customers. They were guaranteed a customer base and a profit by the various state regulatory commissions throughout the United States. The adoption of Mega-NOPR has caused the state legislatures and regulatory agencies to pass laws and rules to deregulate the electric utility industry.

The process of deregulation is progressing rapidly. Many states are passing legislation to break up the utility monopolies. They are trying to encourage competition by allowing end-user's choice of electrical suppliers. Deregulation throughout the United States is expected to be implemented toward the end of this century. For example, in California, the California Public Utilities Commission (CPUC) has proposed to implement deregulation with a phased approach. The CPUC allowed the large industrial end-users to choose their suppliers of electricity on January 2, 1996. The CPUC next plans to allow the following to choose their suppliers of electricity: small industrial end-users in

1997, then commercial customers in 1998, and residential customers in 2002.

In a deregulated environment, utilities will be divided into separate companies. The generation companies will be called *GENCOS*. The transmission companies will be called *TRANSCOS*. The distribution companies will be called *DISCOS*. And the companies providing unbundled energy services will be called *ESCOS*. Most utilities will become TRANSCOS or DISCOS, while the great majority of public utility districts, municipalities, and cooperatives will become DISCOS. The TRANSCOS and DISCOS primary and sometimes only source of revenue will come from GENCOS. GENCOS will pay TRANSCOS and DISCOS for the right to wheel (transit electricity) on their t&d systems. The GENCOS will expect reliable and efficient t&d systems. A reliable and efficient t&d system has many benefits. These benefits include making the TRANSCOS and DISCOS more competitive, reducing the threat of the end-user building its own generation facility, and satisfying regulators that the t&d system is efficient. The TRANSCOS and DISCOS will most likely continue to be regulated monopolies. It is expected that the formation of an independent system operator (ISO) will be necessary to coordinate the operation of the various t&d systems. TRANSCOS and DISCOS probably will find that the regulators will limit how much they can charge for t&d losses. The regulators probably will limit them to an average amount of, say, 3 percent. This means that those parts of a t&d system which exceed the average amount of loss will require efficiency improvements. Otherwise, the TRANSCOS and DISCOS will not recover the cost of the t&d losses that are greater than the average amount. This should provide an incentive for TRANSCOS and DISCOS to improve the efficiency of less efficient parts of their t&d systems, such as low-efficient transformers.

For example, in a nongenerating utility, distribution transformers contribute to almost 40 percent of the losses in the distribution system. If these utilities want to have an efficient distribution system, they must have energy-efficient transformers. Many municipalities, public utility districts, and cooperatives fall into the category of nongenerating utilities. Congress recognized the importance of energy-efficient transformers to the overall efficiency of t&d systems when it passed the 1992 Energy Policy Act.

1992 Energy Policy Act

With the passage of the 1992 Energy Policy Act, the U.S. Congress set in motion the process of deregulating the electric power industry. The purpose of this legislation was to bring competition to an industry dominated for many years by monopolistic vertically integrated utilities. Congress initiated the following three steps to encourage competition in the electric power industry. The first step requires utilities to provide open access to wheel on their t&d systems. The second step requires utilities to separate their power business from their transmission and distribution business. The third step provides end-users with an opportunity to choose the electrical supplier regardless of who provides the transmission and distribution service to the end-user. This last step introduced the idea of retail wheeling, whereby an electrical supplier would wheel power on a local distribution company's system to deliver power to the end-user. This deregulating process is designed to encourage electrical suppliers to compete with one another and supposedly reduce the price of electricity. This deregulation process in the electrical power industry is similar to the deregulation that has already occurred in the telecommunications and airline industries.

The reaction by some utilities to the 1992 Energy Policy Act has been to make decisions regarding additions or improvements to their t&d systems based on short-range factors. Due to uncertainties about the effect of deregulation, these utilities primarily are concerned with keeping capital expenditures to a minimum. They may ignore the long-range benefits of t&d efficiency improvements, such as energy-efficient transformers. This was not the intent of Congress when it passed the Energy Policy Act. The act says in Section III (a) (9):

> The rates charged by any electric utility shall be such that the utility is encouraged to make investments in, and expenditures for, all cost-effective improvements in the energy efficiency of power generation, transmission and distribution.

Efficiency improvements would include energy-efficient transformers. This is reflected in Subtitle C, Section 124 of the Energy Policy Act of 1992, which contains an amendment to Section 346 of the Energy Policy and Conservation Act (42 U.S.C. Section 6317). The following is a portion of that amendment:

> Section 346 (a) (1): The Secretary [Department of Energy] shall, within 30 months after the date of the enactment of the Energy Policy Act of 1992, prescribe testing requirements for those high-intensity discharge lamps and *distribution transformers* for which the Secretary makes a determination that energy conservation standards would be technologically feasible and economically justified, and would result in significant energy savings.
>
> (2) The Secretary shall, within 18 months after the date on which testing requirements are prescribed by the Secretary pursuant to paragraph (1), prescribe, by rule, energy conservation standards for those high-intensity discharge lamps and *distribution transformers*

for which the Secretary prescribed testing requirements under paragraph (1).

In addition to using energy-efficient transformers in new facilities, Congress was interested in the effect of replacing existing transformers with energy-efficient transformers. Consequently, Congress required a study to examine the feasibility of energy savings resulting from transformer replacement. The following excerpt from Section 124 of the 1992 Energy Policy Act states this requirement:

> (c) Study of Utility Distribution Transformers. The Secretary shall evaluate the practicability, cost-effectiveness, and potential energy savings of replacing, or upgrading components of, existing utility distribution transformers during routine maintenance and, not later than 18 months after the date of the enactment of this Act, report the findings of such evaluation to the Congress with recommendations on how such energy savings, if any, could be achieved.

The purpose of the two frequently cited studies by ORNL are to fulfill the requirements of the 1992 Energy Policy Act. The two reports are *The Feasibility of Replacing or Upgrading Utility Distribution Transformers During Routine Maintenance* and *Determination Analysis of Energy Conservation Standards for Distribution Transformers*. These reports contain information to assist the U.S. Department of Energy (DOE) in making a determination on the feasibility and significance of energy conservation for distribution transformers.

The DOE is in the process of developing distribution transformer standards. Within the coming year, DOE will be evaluating alternative transformer test procedures for measuring the efficiency of distribution transformers. The DOE plans to schedule a series of workshops and hearings to gain public input into the adoption of these test procedures. After the test procedures have been adopted, DOE will examine various transformer efficiency standards. Based on this examination, the DOE will specify transformer efficiency standards and schedule another series of public workshops and hearings before adopting these efficiency standards. The rules and regulations for development of these procedures and standards are in the *Federal Register* 61(136), July 15, 1996.

Deregulation versus Regulation

Will the utilities responsible for t&d systems, including transformers, be deregulated or remain regulated? The emphasis up to now in the deregulation process has been to deregulate the generation portion of the utility business. Electric utility industry analysts seem to think that t&d systems or "wires" companies will remain regulated. This need for energy-efficient transformers will continue whether the wires utilities are regulated or deregulated.

If the wires utilities are deregulated, they will have to compete with other wires utilities to participate in the wheeling of power to the end-user of electricity. Utilities that have reliable and efficient transformers will be more competitive in the long run than utilities whose systems are inefficient and unreliable. This was the experience of the telecommunications industry after it was deregulated. In this situation, it will become even more critical for utilities to be concerned about the efficiency of their transformers.

It is more likely that federal and state regulators will continue to regulate the wires utilities. This is so because the wires utilities will continue to be monopolies. If the federal and state regulators regulate the wires utilities, they will insist on the utilities designing their systems to be reliable and efficient. They will either institute incentives to encourage the use of energy-efficient and reliable transformers or set efficiency and reliability standards.

One way the regulators may encourage utilities to use energy-efficient transformers is to put a cap on how much wires utilities can charge for losses. For example, they could limit the wires utilities from using more than 3 percent losses in their wheeling charges. This would cause the wires utilities to bring the rest of their system in compliance with the 3 percent limit or not be able to recover the cost of losses from the GENCOS. The wires utilities would need to use energy-efficient transformers because they realize that distribution transformers are a significant part of their t&d losses.

Another way the regulators will cause the wires utilities to have efficient t&d systems and use energy-efficient transformers is by setting transformer efficiency standards. The 1992 Energy Policy Act requires the DOE to set regulatory efficiency standards for lighting, motors, and *distribution transformers.* In fact, the National Electrical Manufacturing Association (NEMA) has developed the distribution transformer standards. The DOE is accelerating the development of transformer testing requirements and efficiency standards as required under the 1992 Energy Policy Act. Already, the EPA has set efficiency standards for distribution transformers. It has set these standards to reduce emission of carbon dioxide from generators supplying losses to low-efficiency transformers. It has developed the Star Transformer Program. This program is a result of the 1993 U.S. Climate Change Action Plan, which has the goal of encouraging the voluntary reduction of greenhouse gas emissions to 1990 levels by the year 2000. The

EPA requires only voluntary compliance with the Star Transformer standards. If utilities ignore these standards, the regulators might decide to mandate the Star Transformer or some other efficiency standard.

Star Transformer Standards

The Star Transformer standards set up by the EPA are being implemented voluntarily. Utilities that have voluntarily signed on as of May 1996 include Allegheny Power, Mid-Carolina Electric Cooperative, Northeast Utilities Service Company, Oklahoma Gas and Electric, Portland General Electric, Tampa Electric, United Illuminating, and Wisconsin Electric Power. These standards are separate from the standards being developed by the DOE.

The EPA has developed the Star Transformer target efficiency standards based on the values of the A and B factors in the total owning cost formula. Tables 11-1 through 11-5 show these target efficiencies for 10-kVA through 167-kVA single-phase distribution transformers.

The EPA hopes that competition within the electric utility industry will encourage utilities to work with industrial and commercial building customers to share the savings from the use of energy-efficient transformers. In this case, the utility would share the energy savings cost with the customer. The customer has to put up no capital but gets a reduced energy bill.

The EPA has not developed target efficiency standards for substation transformers. This is undoubtedly because substation transformer

Table 11-1. Target Efficiencies for 10-kVA Single-Phase Transformers

	A Factor			
	$00.00–$1.99	$2.00–$3.99	$4.00–$5.99	>$6.00
B factor				
<$0.50	98.17	98.28	98.23	98.28
$0.50–$0.99	98.23	98.65	99.45	98.36
$1.00–$1.49	98.26	98.74	99.90	98.83
$1.50–$1.99	98.38	98.69	99.71	98.81
>$2.00	N/A	98.69	99.93	98.94

NOTE: Target efficiencies are expressed at 50 percent load and may change annually.
SOURCE: Courtesy of EPA.

Table 11-2. Target Efficiencies for 15-kVA Single-Phase Transformers

	A Factor			
	$00.00–$1.99	$2.00–$3.99	$4.00–$5.99	>$6.00
B factor				
<$0.50	98.22	98.33	98.28	98.33
$0.50–$0.99	98.27	98.70	98.50	98.41
$1.00–$1.49	98.31	98.78	98.95	98.88
$1.50–$1.99	98.43	98.73	98.75	98.85
>$2.00	N/A	98.74	98.98	98.99

NOTE: Target efficiencies are expressed at 50 percent load and may change annually.
SOURCE: Courtesy of EPA.

Table 11-3. Target Efficiencies for 25-kVA Single-Phase Transformers

	A Factor			
	$00.00–$1.99	$2.00–$3.99	$4.00–$5.99	>$6.00
B factor				
<$0.50	98.49	98.60	98.55	98.60
$0.50–$0.99	98.54	98.97	98.77	98.68
$1.00–$1.49	98.58	99.05	99.22	99.15
$1.50–$1.99	98.70	99.00	99.02	99.12
>$2.00	N/A	99.01	99.25	99.26

NOTE: Target efficiencies are expressed at 50 percent load and may change annually.
SOURCE: Courtesy of EPA.

Table 11-4. Target Efficiencies for 37.5- to 50-kVA Single-Phase Transformers

	A Factor			
	$00.00–$1.99	$2.00–$3.99	$4.00–$5.99	>$6.00
B factor				
<$0.50	98.64	98.76	98.71	98.75
$0.50–$0.99	98.70	99.13	98.92	98.83
$1.00–$1.49	98.73	99.21	99.38	99.31
$1.50–$1.99	98.86	99.16	99.18	99.28
>$2.00	N/A	99.17	99.41	99.42

NOTE: Target efficiencies are expressed at 50 percent load and may change annually.
SOURCE: Courtesy of EPA.

Table 11-5. Target Efficiencies for 75- to 167-kVA Single-Phase Transformers

	A Factor			
	$00.00–$1.99	$2.00–$3.99	$4.00–$5.99	>$6.00
B factor				
<$0.50	98.98	99.05	98.90	99.26
$0.50–$0.99	99.02	99.25	99.17	99.01
$1.00–$1.49	99.19	99.29	99.32	99.46
$1.50–$1.99	99.11	99.21	99.34	99.40
>$2.00	N/A	99.26	99.36	99.59

NOTE: Target efficiencies are expressed at 50 percent load and may change annually.
SOURCE: Courtesy of EPA.

losses contribute a much smaller amount to greenhouse gas emissions than distribution transformer losses. Nevertheless, similar standards are needed for substation transformers. Why should they be exempt from reducing greenhouse gas emissions?

In order to facilitate the use of energy Star Transformers, EPA has developed three computer programs. They are the Commercial and Industrial Transformer Cost-Evaluation Model (CITCEM), the Transformer Sizing Program (TSP), and the Distribution Transformer Cost-Evaluation Model (DTCEM). The CITCEM computer program is designed to help facility managers, utilities, and energy consultants evaluate the cost-effectiveness of the application of energy-efficient dry-type distribution transformers. The CITCEM computer program is expected to be released in 1997. The TSP computer program is being developed in collaboration with the Edison Electric Institute and will help utility engineers and field personnel determine the optimal transformer rating under various operating conditions. The TSP computer program will help utilities and commercial/industrial users of transformers save energy and money by sizing transformers properly. The DTCEM computer program was released in December of 1996 and is included with this book. The DTCEM program is designed to assist utility engineers and technicians, particularly those with smaller utilities, to evaluate the cost-effectiveness of energy-efficient transformers.

EPA's DTCEM Computer Program

The EPA has developed the DTCEM computer program to help utilities to implement the EPA's transformer efficiency standards. This computer program provides a step-by-step procedure for determining the A and B factors that are used in the total owning cost formula. The same methods described earlier in the book are applied in this computer program. The computer program uses the load characteristics to determine the energy transformer losses in kilowatt hours. Load characteristics include hours per year the transformer operates, the equivalent annual peak load, the transformer loss factor, and peak responsibility factor. The cost of providing and transmitting electric capacity and energy for transformer losses depends on the avoided cost to generate and transmit the necessary energy and capacity. The program calculates the values of the A and B factors from the avoided cost to generate and transmit the transformer losses. The program provides a tutorial containing a case study of a typical utility evaluating six transformers. It takes the user through the following steps:

1. Enter utility information including the name of the utility, its location, and amount of CO_2, SO_2, and NO_x emissions the utility produces in pounds per kWh.

2. Enter financial factors including fixed charge rate, base year, the number of years in the evaluation, annual inflation rate, and discount rate.

3. Choose from three alternative methods for calculating the A and B factors: for a generating utility based on generating costs; if already known, enter directly; and for a nongenerating utility (DISCO) based on the demand and energy charges.

4. Enter the transformer type and size.

5. Enter transformer load characteristics.

6. Enter cost factors including the cost of generation capacity and energy and cost of transmission and distribution capacity.

7. Enter the cost of the transformer being considered based on the bids.

The computer program will then calculate the total owning costs of each of the transformers being evaluated and print out a bid evaluation report. It also will compare the efficiency values and determine if they comply with the EPA's energy star standards. It also will calculate the simple payback by dividing the price difference of the transformers being evaluated by the energy savings per year.

The floppy disk of the computer program is in the back of the book. The appendix contains the user's manual for EPA's DTCEM computer program.

NEMA Efficiency Standards

NEMA also has developed minimum energy efficiency standards for liquid-filled and dry-type distribution transformers. These standards are for utilities and commercial/industrial users of transformers who do not evaluate transformer losses. They are contained in the NEMA Standards Publication No. TP-1, "Guide for Determining Energy Efficiency for Distribution Transformers," 1300 N. 17th Street, Rossyln, VA 22209.

Research and Development

Research and development (R&D) in the electric utility industry has shifted from technologies that meet regulators' requirements to technologies that will help utilities compete in the deregulated market. This shift in emphasis should not affect R&D for energy-efficient transformers. R&D for improving the efficiency of transformers needs to continue. The wires utilities that will be purchasing and installing transformers will most likely continue to be regulated. There will be an increasing need for industrial and commercial users of transformers to purchase more efficient transformers. The state and federal regulators will continue to set standards and regulations that will encourage the use of energy-efficient transformers. In the past, EPRI collaborated with the transformer and utility industry and supported the development of technologies for improving the efficiency of transformers. This research has resulted in the commercialization of amorphous steel core distribution transformers and the development of the ribbon cable advanced transformer design. Research needs to continue in expanding amorphous steel cores to substation transformers and bringing down the cost of producing and using amorphous steel in distribution transformers. Research into new insulation, such as aramid insulation, needs to be expanded. New cooling media need to be explored, such as the recently developed Edisol TR.

Edisol TR is a synthetic that has superior qualities over transformer oil. It allows the use of a smaller transformer at the same or greater efficiency rating. This synthetic fluid has two main components: polyaphaolefin (PA) and phenylorthoxylyethane (POXE). This synthetic has a greater ability than conventional oil to withstand electric stress. Consequently, its use in transformers results in nearly a 60 percent reduction in fluid volume over conventional oil and a weight decrease of nearly 30 percent. Even with the increased amount of core and coil material needed to improve transformer efficiency, energy-

Table 11-6. Volume, Weight, and Efficiency for One-Phase, 25-kVA Transformers

Characteristics	Nonevaluated	Evaluated ($3/$1)*	Nonevaluated new design
Weight	176.9 kg	156.0 kg	121.5 kg
Height	114.3 cm	94.0 cm	88.9 cm
Fluid	84.5 liter	42.1 liter	34.1 liter
No-load loss	77 W	56 W	73 W
Load loss	526 W	314 W	555 W
Efficiency@50%	98.4%	98.9%	98.3%

*Capitalized cost of loses in U.S. dollars (no-load/load).

SOURCE: Gonzalea, D. A., G. Gauger, A. Yerges, and G. Goedde, Distribution Transformer for the 21st Century, '97 CIRED Conference.

efficient distribution transformers with Edisol TR rather than conventional transformer oil are smaller and weigh less. Table 11-6 illustrates this point.

Superconducting Transformers

Superconducting transformers became a viable alternative to other energy-efficiency transformer technologies with the development of high-temperature superconductors in 1986. High-temperature superconductors (HTS) are resistance-free conductors made of ceramic materials that have superconducting characteristics up to $-171°F$, while low-temperature superconductors were discovered in 1911 and exhibit superconducting properties at $-452°F$. Both high-temperature and low-temperature superconductors require cryogenic equipment to cool the conductors to these very low temperatures. High-temperature superconductors require less energy than low-temperature superconductors to cool the conductors. High-temperature superconductors can use liquid nitrogen as a cryogen to cool the HTS wires. Liquid nitrogen is inexpensive, inert, and nontoxic and is formed into gaseous nitrogen when chilled to $-320°F$. Superconducting technology applied to transformers would reduce the resistance or load losses to zero. The application of this technology has been hampered by the brittle properties of ceramic superconductors. Recently, these barriers have been overcome with the installation of a high-temperature superconducting transformer on a utility's system.

In March of 1997, Service Industriels de Geneve installed the first demonstration of an HTS transformer in the Geneva Electric Power

Grid. The company installed it to demonstrate the application of HTS transformers on utility systems. Asea Brown Boveri (ABB) built the HTS transformer. American Superconductor Corporation provided the superconductors for the HTS transformer. The 630-kVA, 18.7- to 420-kV transformer is a prototype of HTS transformers that are expected to be commercially available by the year 2000. It uses liquid nitrogen rather than oil as a coolant and is half the size and weight of a conventional transformer. Unlike oil, the liquid nitrogen coolant is nontoxic, environmentally friendly, and nonflammable.

Basic Formulas

P = present worth

F = future worth

A = annual cash flow

n = number of years

i = interest rate in percent per year

a = inflation rate in percent per period

CRF_n = uniform series capital recovery factor at interest i for n years

Term	Symbols	Given	Find	Formula
Compound interest factor	CIF	P	F	$(1+i)^n$
Single present-worth factor	PWF	F	P	$\dfrac{1}{(1+i)^n}$
Uniform series PW factor	USPWF	A	P	$\dfrac{(1+i)^n-1}{i(1+i)^n}$
Capital recovery factor	CRF	P	A	$\dfrac{i(1+i)^n}{(1+i)^n-1}$
Compound amount factor	CAF	A	F	$\dfrac{(1+i)^n-1}{i}$
Sinking fund factor	SFF	F	A	$\dfrac{i}{(1+i)^n-1}$
PW of inflation series	PWIS	A	P	$\dfrac{1-\left(\dfrac{1+a}{1+i}\right)^n}{(i-a)}$ for $a{\neq}i$ $\dfrac{n}{1+i}$ for $a = i$

Minimum Acceptable Return (*MAR*)

Minimum acceptable return is defined as the opportunity cost of capital for the transformer purchaser. It is also referred to as the *cost of capital, discount rate* or *interest rate.*

The following is an example of how to calculate the *MAR*. A utility has the following debt ratio, debt cost, stock cost, stock ratio, and stock cost. What is the *MAR?*

solution

Investment components	Rate
Debt ratio	40 percent
Debt cost	7 percent/year
Preferred stock ratio	10 percent
Preferred stock cost	8 percent
Common stock ratio	30 percent
Common stock cost	13 percent

Determining the interest rate *I* or the rate of return requires several calculations. The rate of return is the weighted cost of capital including inflation.

$$MAR = \text{(debt ratio} \times \text{debt cost)}$$
$$+ \text{(preferred stock ratio} \times \text{preferred stock cost)}$$
$$+ \text{(common stock ratio} \times \text{common stock cost)}$$

Minimum acceptable rate of return I is calculated as follows:

$$MAR = (0.40 \times 0.070) + (0.10 \times 0.08) + (0.30 \times 0.13) = 0.075$$

MAR is important because it is used in all the transformer economics methods presented in this book. It provides a basis for accounting for the time value of money.

Appendix C

Carrying Charge Rate or Fixed Charge Rate (*FCR*)

Carrying charge rate or *fixed charge rate* was discussed briefly in Chap. 4. This term converts the levelized annual cost of losses into a capitalized value. It is often referred to as the *annual cost ratio*. It can be multiplied by the bid price to convert the cost of the transformer into an annual cost. Or it can be divided into the annual cost of transformer losses to convert the annual cost of transformer losses into a capitalized cost. Four components make up the carrying charge rate:

1. Minimum acceptable return

2. Book depreciation

3. Income taxes

4. Local property taxes and insurance

Example

Components	Investor owned	Municipals, cooperatives
Minimum acceptable return	0.100	0.07
Book depreciation	0.015	0.01
Income taxes	0.015	0.00
Local property taxes and insurance	0.030	0.03
Fixed charge rate (total)	0.160	0.10

Simple Interest Rate

Simple interest rate is the interest paid for a loan and is directly proportional to the capital involved in the loan. Mathematically, the total paid for interest I can be illustrated by the following formula:

$$I = PiN$$

where P = present amount or principal
  i = interest rate per period of study
  N = number of interest periods

Since the principal borrowed P is a fixed value, the annual interest is constant. The total amount a borrower is obligated to pay is as follows:

$$F = P + I = P + PiN = P(1 + iN)$$

The following is an example of how to calculate the interest owed from a loan. A commercial office complex must borrow $100,000 at 7 percent interest and pay off the loan in 10 years to purchase $100,000 worth of transformers. What is the interest owed after 10 years?

Solution The interest owed is calculated as follows:

$$I = PiN = (100000)(0.07)(10) = \$70,000$$

Total amount = $100,000 + $70,000 = $170,000

Appendix **E**

Cash Flow Diagram

A useful tool in performing transformer economic analysis is the *cash flow diagram*. The cash flow diagram is a graph of cash flow (in the vertical direction) as a function of time (in the horizontal direction). An arrow pointing up represents expense, while an arrow pointing down represents income or savings. Figure E-1 illustrates the application of a cash flow diagram to a transformer purchase.

The cash flow diagram provides a visual means to sort out the flow of money associated with the purchase of a transformer. It helps provide a visual means to illustrate the various methods for evaluating transformers, starting with present-worth analysis.

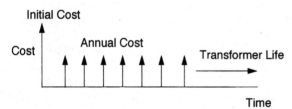

Figure E-1. Cash flow diagram.

Appendix F

Present-Worth Analysis

The present-worth concept is based on the fact that money today has a different value than it will have in the future. It is like a person in a time machine that will only go forward into the future. The time traveler cannot change what has happened in the past but can only change the future. This is the way it is in real life. Past decisions have an effect on today's situation. But past decisions cannot be changed. Getting back to our time traveler. This person has $231.30 in his wallet when he leaves in 1996 and is transported to the year 2006. First thing the time traveler will notice when he arrives in the year 2006 is that he has more money in his wallet. He finds he has $600 in his wallet. What happened? Could he have predicted how much money he would have when he arrived in the year 2006? He could have if his on-board computer had been programmed to calculate present- and future-worth values. There are two present-worth analysis factors: single present-worth (*SPW*) factor and uniform series present-worth (*USPW*) factor. The one factor the time traveler would need is the single present-worth factor. This factor is described by the following formula, where the interest rate is *i* and the number of periods is *n*:

$$SPW = \frac{1}{(1 + i)^n}$$

where n = number of periods
i = the minimum rate of return or discount factor in percent per year

It is used to determine the present worth of a future amount F discounted to the present worth P and is described by the following formula:

$$P = SPW \times F$$

It can be illustrated by the cash flow diagram in Fig. F-1.

How to use SPW factor can best be illustrated by the following problem:

How much money would you have today to deposit in an interest-bearing account at 10 percent a year to have enough money to replace a 25-kVA transformer costing \$600 in 10 years? See Fig. F-2.

Solution

$$P = SPW \times F = \frac{1}{(1 + 0.1)^{10}} \times \$600 = (0.3885) \times \$600 = \$231.30$$

The second type of present-worth analysis deals with converting equal annual costs into present worth. The single-payment present-worth method requires multiplying the present-worth factor for each year by the annual cost for that year. There is a factor called the *present-worth factor for a uniform series* that allows you to multiply the annual cost by the uniform series present-worth factor to obtain the equivalent present worth of the annual cost. This is when a time traveler knows the annual cost to maintain his or her time machine but does not want to have to go to every year and back to the present to find out the total present-worth cost of maintaining his or her time machine. All he or she needs to do is program his or her computer to calculate the uniform series present-worth ($USPW$).

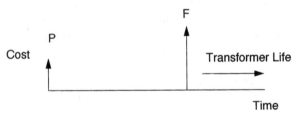

Figure F-1. Single-payment present-worth cash flow diagram.

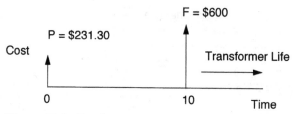

Figure F-2. Single-payment present-worth example cash flow diagram.

$$USPW = \frac{(1 + i)^n - 1}{i(1 + i)^n}$$

where n = number of periods (years)
i = the minimum rate of return or discount factor in percent per year

It is used to determine the present worth of a future amount F discounted to the present worth P and is described by the following formula:

$$P = USPW \times A$$

It can be illustrated by the cash flow diagram in Fig. F-3.

How to use the $USPW$ factor can best be illustrated by the following problem:

How much money would the transformer loss savings be worth today, if through the use of a 25-kVA amorphous steel core transformer

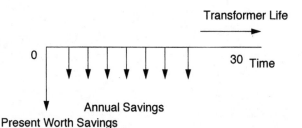

Figure F-3. Uniform series payment present-worth cash flow diagram.

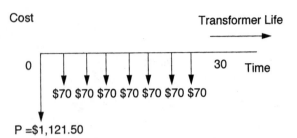

Figure F-4. Uniform series payment present-worth example cash flow diagram.

$70 per year of no-load losses could be saved over the 30-year life of the transformer? See Fig. F-4.

Solution

$$P = USPW \times A = \frac{(1 + 0.1)^{30} - 1}{0.1(1 + 0.1)^{30}} \times \$70$$

$$= (17.45) \times \$70 = \$1221.50$$

Appendix **G**

Compound Recovery Factor (*CRF*)

The *compound recovery factor* concept is based on the fact that money today can be spread out in equal payments over time. It is the reciprocal of the uniform series present-worth (*USPW*) factor. It is a method for determining the present worth of a series of equal payments. This factor is described by the following formula, where the interest rate is i and the number of periods is n:

$$CRF = \frac{i(1 + i)^n}{(1 + i)^n - 1}$$

where n = number of periods
$\quad i$ = the minimum rate of return or discount factor in percent per year

It is used to determine the present worth P of a series of payments A and is described by the following formula:

$$P = CRF \times A$$

It can be illustrated by the cash flow diagram in Fig. G-1.

How to use *CRF* can best be illustrated by the following problem:

How much money would be saved per year at 10 percent interest for 30 years to justify spending \$250 more for a high-efficiency 25-kVA transformer? See Fig. G-2.

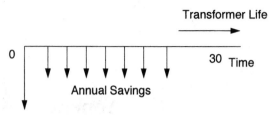

Present Worth Savings

Figure G-1. Compound recovery cash flow diagram.

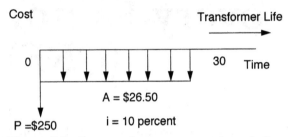

Figure G-2. Compound recovery example cash flow diagram.

Solution

$$A = CRF \times P = \frac{0.1(1 + 0.1)^{30}}{(1 + 0.1)^{30} - 1} \times \$250 = 0.106 \times \$250 = \$26.50$$

Compound Amount Factor (*CAF*)

The *compound amount factor* concept is based on the fact that money spread out over a period of time can be accumulated into a future value. It is a method for determining the future value of a series of equal payments. This factor is described by the following formula, where the interest rate is *i* and the number of periods is *n:*

$$CAF = \frac{(1 + i)^n - 1}{i}$$

where n = number of periods

i = the minimum rate of return or discount factor in percent per year

It is used to determine the future amount *F* of a series of payments *A* and is described by the following formula:

$$F = CAF \times A$$

It can be illustrated by the cash flow diagram in Fig. H-1.

How to use *CAF* can best be illustrated by the following problem:

How much money would $70 per year of lost savings be worth in 30 years at 10 percent interest due to the purchase of a high-efficiency 25-kVA transformer? See Fig. H-2.

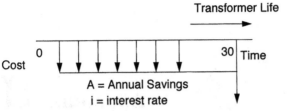

Figure H-1. Compound amount factor cash flow diagram.

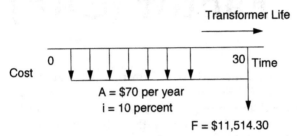

Figure H-2. Compound amount factor example cash flow diagram.

Solution

$$F = CAF \times A = \frac{(1 + 0.1)^{30} - 1}{0.1} \times \$70 = 164.49 \times \$70 = \$11{,}514.30$$

Appendix I
Sinking-Fund Factor (*SFF*)

The *sinking-fund factor* concept is a factor for determining how much money must be saved each year to obtain a certain future amount. It is a method for determining the series of equal payments necessary to achieve a future worth. It is the reciprocal of the capital recovery factor. This factor is described by the following formula, where the interest rate is i and the number of periods is n:

$$SFF = \frac{i}{(1 + i)^n - 1}$$

where n = number of periods
i = the minimum rate of return or discount factor in percent per year

It is used to determine the future amount F of a series of payments A and is described by the following formula:

$$A = SFF \times F$$

It can be illustrated by the cash flow diagram in Fig. I-1.

How to use the *SFF* can best be illustrated by the following problem:

How much money would be needed to save for 30 years at 10 percent interest in order to have the extra \$250 to purchase a high-efficiency 25-kVA transformer? See Fig. I-2.

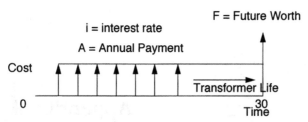

Figure I-1. Sinking-fund factor cash flow diagram.

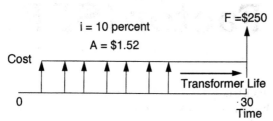

Figure I-2. Sinking-fund factor example cash flow diagram.

Solution

$$A = SFF \times F = \frac{0.1}{(1 + 0.1)^{30} - 1} \times \$250 = 0.00608 \times \$250 = \$1.52$$

Appendix J

Present Worth of Inflation Series

In order to evaluate the value of losses, it is necessary to escalate the cost of fuel and determine its present worth. This is done by use of the *present worth of inflation series* (*PWIS*). This factor is described by the following formula, where the interest rate is i, the inflation rate is a, and the number of periods is n:

$$PWIS = \frac{1 - [(1 + a)/(1 + i)]^n}{i - a} \qquad \text{for } a \neq i$$

It is used to determine the present worth P of a series of payments A escalating over time and is described by the following formula:

$$P = PWIS \times Ai$$

where Ai = first-year value of losses

It can be illustrated by the cash flow diagram in Fig. J-1.

How to use *PWIS* can best be illustrated by the following problem:

How much would the value of losses be worth today for 30 years at 10 percent interest, inflation rate of 4.5 percent, if the first year value of losses is 2.5 cents/kWh? See Fig. J-2.

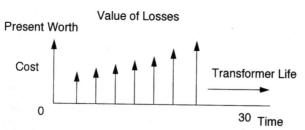

Figure J-1. Present worth of inflation series cash flow diagram.

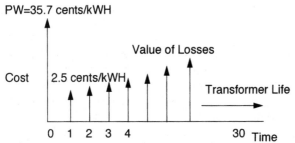

Figure J-2. Present worth of inflation series example cash flow diagram.

Solution

$$P = PWIS \times Ai$$

$$= \left[\frac{1 - [(1 + 0.045)/(1 + 0.1)]^{30}}{0.1 - 0.045} \right] (2.5 \text{ cents/kWh})$$

$$= 14.28(2.5 \text{ cents/kWh})$$

$$= 35.70 \text{ cents/kWh}$$

Levelizing

Levelizing is the conversion of unequal annual payments to uniform annual payments. In this method, the first step is to convert each unequal annual payment to its present worth. The next step is to multiply the present-worth value by the capital recovery factor for the life of the study. For example, the product of the present-worth inflation series (*PWIS*) and the capital recovery factor for n years (CRF_n) is often referred to in the utility industry as the escalation factor (*EF*):

$$EF = \left[\frac{1 - [(1 + a)(1 + i)]^n}{i - a} \right] CRF_n \qquad \text{for } a \neq i$$

$$EF = \left(\frac{n}{1 + i} \right) CRF_n \qquad \text{for } a = i$$

Then

$$\text{Levelized energy cost} = EF \times Ai$$

where Ai = first-year value of losses

The same problem as in App. J can be used to illustrate how to convert the value of losses, i.e., avoided fuel cost, to equal annual amounts:

How much would the value of losses be worth each year for 30 years at 10 percent interest, inflation rate of 4.5 percent, if the first year value of losses is 2.5 cents/kWh? See Fig. K-1.

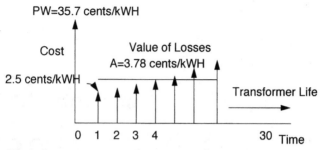

Figure K-1. Levelizing example cash flow diagram.

Solution

Levelized energy cost

$$= EF \times 2.5 \text{ cents/kWh}$$

$$= \left[\frac{1 - [(1 + 0.045)/(1 + 0.1)]^{30}}{0.1 - 0.45} \right] \left[\frac{0.01\,(1 + 0.1)^{30}}{(1 + 0.1)^{30} - 1} \right] (2.5 \text{ cents/kWh})$$

$$= (14.28)\,(0.106)(2.5 \text{ cents/kWh})$$

$$= 3.78 \text{ cents/kWh}$$

Appendix **L**

DTCEM User's Manual

A Manual for Use with the Distribution Transformer Cost Evaluation Model (DTCEM) (Version 1.1; April 1997)

This manual is to be used in conjunction with the DTCEM program disk, which appears inside the back cover of this book.

This material was prepared by the U.S. Environmental Protection Agency (EPA). Copyright © 1997 by the U.S. Environmental Protection Agency. Used by permission of the authors and the publisher. Appreciation is acknowledged to Gregory L. Booth, PE, RLS of Booth & Associates, for his contribution to formulas and tables in this book.

Introduction

The goal of EPA's Energy Star Transformer Program is to encourage electric utilities to purchase and install high-efficiency distribution transformers where they are cost-effective. Another goal of the Program is to provide state-of-the-art technical tools which will assist utility efforts to cost-effectively accelerate pollution prevention.

The Distribution Transformer Cost Evaluation Model (DTCEM) helps electric utilities, particularly smaller utilities, perform the complex economic analyses needed to accurately determine the cost-effectiveness and emission reduction potential of high-efficiency distribution transformers. DTCEM provides the information necessary for utilities to weigh purchases of high-efficiency distribution transformers against other competing resource options.

Installing the DTCEM Software

Before you begin working with DTCEM, check the contents of your DTCEM package, make sure you have the correct equipment to run the program, and read through the rest of this section to be sure you have a clear understanding of the installation procedure.

The DTCEM Package

Your DTCEM package includes the following a 3½ inch DTCEM program disk and this user's manual.

Required Equipment

- An IBM compatible computer with a 386SX or better processor with 4MB RAM;
- Microsoft Windows 3.1 or later; and
- Hard disk with at least 4 MB of space available.

Recommended Equipment

- *Color monitor* - DTCEM operates on a monochrome monitor; however, some screens are difficult to read. We suggest using a screen resolution greater than 640 x 480 (at this resolution, some parts of the screens may be cut off).
- *Mouse* - If you do not have a mouse, it is possible (though rather inconvenient) to use DTCEM using keyboard controls. File menu options may be accessed by clicking the Alt key and the underscored letter in the menu option (e.g., to access the **File** menu, click **Alt+F**).
- *Printer* - You may wish to print a hard copy of DTCEM's results.

Installation Instructions

To install DTCEM on your computer, follow the instructions below:

1. Insert the DTCEM disk into your floppy disk drive (A or B).
2. Click on the **File** menu of your **Windows Program Manager** and select **Run**.
3. Type **a:\install** (or **b:\install**) and click **OK**.
4. Follow the instructions during the installation process, *making sure that you select the default directory.* **DTCEM will not be installed correctly if you modify the default drive or directory.**
5. Read the message in the instruction screen at the end of the installation process and double click on the upper left hand corner to continue.

To run DTCEM, double click on the DTCEM icon, or click on the File menu of the Windows Program Manager, select Run, and type c:\dtcem\dtcem.exe (or d:\dtcem\dtcem.exe or e:\dtcem\dtcem.exe, depending on where you install the DTCEM program files).

After DTCEM has loaded, it will display the "Welcome to DTCEM" screen (Figure 1). To begin the program, click on **Yes**.

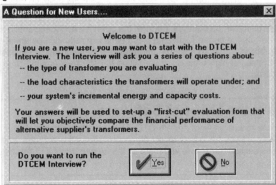

Figure 1: DTCEM welcome screen

If you are a first time user, you may want to continue with the DTCEM Tutorial section of this manual (Chapter 1) to go through a quick overview of the DTCEM software.

If you have any questions regarding the above installation procedure, please call the Energy Star Hotline at **1-888-STAR-YES** (toll-free).

Organization of this Manual

This manual is divided into chapters which describe the different concepts within the DTCEM software program. The chapters are organized as shown below:

Chapter 1. DTCEM Interview and Tutorial. This chapter introduces the DTCEM Interview and presents a case study showing how sample data may be entered and analyzed for potential high-efficiency distribution transformer purchases.

Chapter 2. Setting Up Defaults. This chapter describes how to enter and edit the default data required for DTCEM. This data includes the utility name, state, emissions factors, financial data, and cost estimates.

Chapter 3. Bid Evaluation Screen. This chapter describes how to enter the information required to evaluate transformer bid evaluations.

Chapter 4. DTCEM Toolbox. This chapter describes how to edit the default data used in the Bid Evaluation.

Chapter 5. Help and Other Features. This chapter describes the general features of DTCEM including the online help system and general windows attributes.

Figure 2 outlines the organization of the DTCEM screens based on these chapters. Gray boxes represent default screens defined in Chapter 2 (Setting Up Defaults); white boxes represent screens defined in Chapter 3 (The Bid Evaluation Screen); and black boxes represent screens defined in Chapter 4 (DTCEM Toolbox). In addition, boxes corresponding to screens described in Chapter 1 (DTCEM Interview and Tutorial) are indicated by circled numbers which corresponds to the Step described in the DTCEM Interview.

Figure 2: Organization of DTCEM Screens

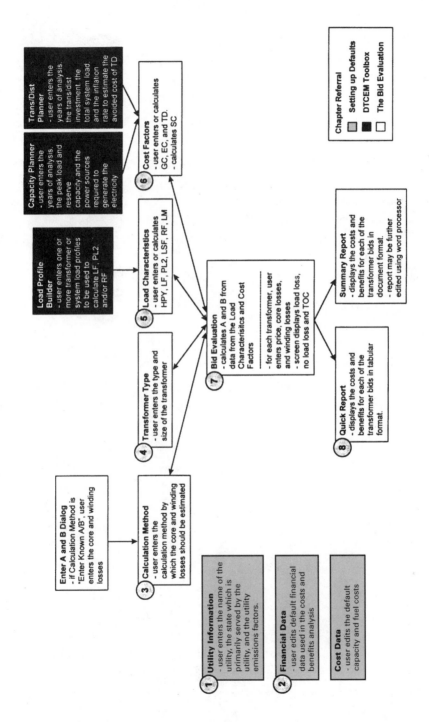

Trans/Dist Planner
- user enters years of analysis, the trans/dist investment, the total system load, and the inflation rate to estimate the avoided cost of TD

Capacity Planner
- user enters the years of analysis, the peak load and reserve capacity, and the power sources required to generate the electricity

Load Profile Builder
- user enters one or more transformer or system load profiles to be used to calculate LF, PL2 and/or RF

6 **Cost Factors**
- user enters or calculates GC, EC, and TD.
- calculates SC

5 **Load Characteristics**
- user enters or calculates HPY, LF, PL2, LSF, RF, LM

4 **Transformer Type**
- user enters the type and size of the transformer

7 **Bid Evaluation**
- calculates A and B from data from the Load Characterisitcs and Cost Factors
- for each transformer, user enters price, core losses, and winding losses
- screen displays load loss, no load loss and TOC

Enter A and B Dialog
- if Calculation Method is "Enter Known A/B", user enters the core and winding losses

3 **Calculation Method**
- user enters the calculation method by which the core and winding losses should be estimated

Summary Report
- displays the costs and benefits for each of the transformer bids in document format.
- report may be further edited using word processor

Quick Report
- displays the costs and benefits for each of the transformer bids in tabular format.

8

1 **Utility Information**
- user enters the name of the utility, the state which is primarily served by the utility, and the utility emissions factors.

2 **Financial Data**
- user edits default financial data used in the costs and benefits analysis

Cost Data
- user edits the default capacity and fuel costs

Chapter Referral

Setting up Defaults

DTCEM Toolbox

The Bid Evaluation

203

Chapter 1. DTCEM Interview and Tutorial

This tutorial is intended for those who are not familiar with DTCEM. Following a brief primer describing a common method for evaluating transformer bids, a case study is presented showing how sample data may be entered and analyzed for potential high-efficiency distribution transformer purchases.

Primer - How to Evaluate a Transformer Bid.

Transformers are used to step down voltages. Because transformers are not 100 percent efficient, transformer losses represent a continuous source of losses to a utility or cooperative. The financial value of these losses depends upon the loss characteristics of a transformer, the characteristics of the load it is operated under, and the cost of providing electric capacity and energy. The cost-effectiveness of a new transformer purchase depends upon not only its purchase price but also the cost of operating it over its lifetime. DTCEM allows utilities and cooperatives to easily evaluate the cost-effectiveness of new transformer purchases.

Transformers exhibit two types of losses: core losses and winding losses. Core losses occur when the transformer is energized, even if there is no load. These losses occur because of the electrical currents and magnetic fields that magnetize the transformer core. Core losses are constant regardless of the transformer load. Winding losses, on the other hand, occur only when the transformer is loaded. Winding losses are due to the normal I^2R losses exhibited by any standard conducting material. Winding losses vary with the square of the load on the transformer. Note that core losses are also referred to as no-load losses and that winding losses are also referred to as load losses.

When a transformer bid is received, the bid typically specifies the transformer's purchase price, its core losses (in watts), and its winding losses (in watts). The basic method of transformer evaluation (for investor owned utilities) is to assign a dollar cost per watt of core losses and a dollar cost per watt of winding losses. The dollar value assigned to a watt of core losses is often referred to as the 'A' factor and the dollar value assigned to a watt of winding losses is often referred to as the 'B' factor. The overall cost of a transformer can then be expressed as:

Total Owning Cost = Purchase Price + (A * Core Losses) + (B * Winding Losses)

where: A = present value cost of a watt or core losses ($/watt)

                B = present value cost of a watt of winding losses ($/watt)

The transformer purchasing process generally works as follows. After estimating the utility specific A and B factors (described below), a utility or cooperative solicits bids from interested manufacturers. The solicitation includes the specific technical specifications required for the transformers as well as the utility's or cooperative's specific A and B factors. The prospective transformer suppliers respond by proposing their transformers that best meet the utility's or cooperative's requirements. After receiving the transformer bids from each manufacturer and checking that they meet the minimum technical specifications required (and any other requirements), each manufacturer's transformer bid can be evaluated using the formula above and the transformer with the lowest total owning cost identified. As we will see in the tutorial to follow, the transformer with the lowest overall cost is not necessarily the transformer with the lowest purchase price.

In order to calculate the A and B factors, information is required on both the load characteristics the transformers will operate under and the cost of providing electrical capacity and energy. For load characteristics, this includes the hours per year the transformer operates (HPY), the equivalent annual peak load (PL), the transformer loss Factor (LsF), and the peak responsibility factor (RF). The cost of providing electric capacity and energy depends upon the costs avoided by not having to provide a kW of capacity and by not generating a kWh of energy.

Case study:

ABC Utility is an investor owned utility providing power primarily to the state of Indiana. ABC is looking to purchase 500 single phase oil filled transformers rated at 75 kVA, 120/240 voltage. Use the DTCEM tutorial to calculate ABC Utility's core and winding losses (A and B factors). Using the sample set of transformers bids, the tutorial will also show you how to identify the most cost effective transformer.

Assume that ABC received the following transformer bids.

Manufacturer	Bid	No-load (Core) loss	Load (Winding) loss
Manufacturer 1	$390	95 watts	413 watts
Manufacturer 2	$375	115 watts	390 watts
Manufacturer 3	$407	99 watts	385 watts
Manufacturer 4	$420	104 watts	365 watts
Manufacturer 5	$388	107 watts	410 watts
Manufacturer 6	$412	90 watts	430 watts

Start this case study by installing and running DTCEM as explained in the Installation Instructions on page 1. This case study describes entering the data through the DTCEM Interview. Upon starting the DTCEM Program you may start the interview by clicking "**Yes**" in the Welcome box. You may also start the interview by selecting **Interview** from the **Interview** menu.

The first interview screen pops up as shown in Figure 2 below:

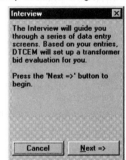

Figure 2: Interview Explanation screen

→   *To continue with the interview, click on the Next button.*

The steps involved with entering the data into DTCEM are outlined below. Interview help for each step is displayed in the upper or lower right hand corner of the screen.

Step 1. Enter Utility Information

The first step is to enter the information about ABC Utility.

→   *Enter "ABC Utility" in the utility name box and select "Indiana" from the state pick list.*

For now, accept the default values for the emissions factors listed in the lower portion of the box (you may change these values by clicking in the appropriate white boxes, deleting the current values, and typing the correct values). Your finished "Utility Information" screen should look like Figure 3 below:

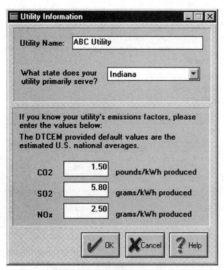

Figure 3: ABC Utility Information

→   **When you are finished, click on the Next button to continue with the Interview.**

Step 2. Enter Financial Factors

The next step is to enter the financial factors which will be used in the cost and benefit calculations. For this case study we will accept the default values given in this screen for all of the financial factors. The "Enter Financial Factors" screen looks like Figure 4 below:

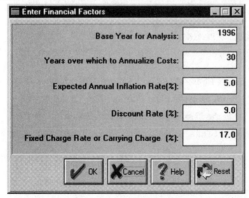

Figure 4: Financial Factors screen

This screen contains the base year for analysis, the years over which to annualize costs, the expected annual inflation rate (page 58), the discount rate (page 58), and the fixed charge rate (page 58).

→   **Accept the defaults for these financial factors by clicking on the Next button.**

Step 3. Select a Calculation Method

The next step is to select the calculation method for estimating core and winding losses (A and B factors). You have the following three choices in this "Select the Calculation Method" screen (Figure 5):

1. Enter Losses Directly
2. Calc Losses (Generator)
3. Calc Losses (Disco)

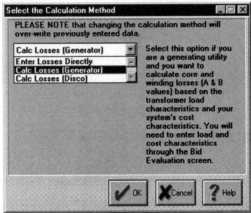

Figure 5: Calculation Method dialog box

→ *Since ABC Utility is a generating utility and we want to calculate the core and winding losses, select "Calc Losses (Generator)" from the drop down list.*

→ *Click on the Next button to continue with the interview.*

Step 4. Enter Transformer Type and Size

The next step is to enter the size and type of transformers which ABC Utility wishes to purchase.

→ *Select Single Phase Oil Filled from the "Transformer Type" picklist.*

→ *Select 75 kVA from the "Transformer Size" picklist.*

→ *Select 30 in the "Transformer Lifetime" picklist.*

The completed "Select Transformer Type and Size" screen should look like Figure 6 below:

Figure 6: ABC Transformer Type and Size Dialog Box

→   **Click on the Next button to continue with the interview.**

Step 5. Enter Load Characteristics

The next step is to enter the load characteristics, specific to the transformer's operating load. The transformer's operating load characteristics are used to estimate the amount of energy that will be lost by the transformer over its lifetime. The "Load Characteristics" screen is shown in Figure 7.

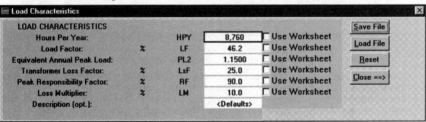

Figure 7: Load Characteristics screen

This screen includes default values for the load characteristics including:

- Hours Per Year (page 58)
- Load Factor (page 59)
- Equivalent Annual Peak Load (page 58)
- Transformer Loss Factor (page 60)
- Peak Responsibility Factor (page 59)
- Loss Multiplier (page 58)

These factors are used in the calculation of the load and no-load losses (except for the load factor which is only used for reporting purposes only). The values may be entered in this screen in one of two ways. First, the values may be directly entered by clicking in the corresponding cream colored cell and typing the correct value. Alternatively, the values may be calculated by clicking on the corresponding "Use Worksheet" box to the right of the value and entering the information needed to calculate the value in the calculation sheet at the bottom of the screen.

For ABC Utility we will calculate the value for the Equivalent Annual Peak Load (PL) and
accept the defaults for the remaining factors.

→   *Click in the "Use Worksheets" box next to the "Equivalent Annual Peak
     Load" cell.*

The Load Characteristics screen expands to show a calculation sheet with six tabs in the
lower portion of the screen as shown in Figure 8. Only tabs corresponding to clicked "Use
Worksheets" boxes in the upper portion of the screen are active. In this case, only the
"Equiv. Annual Peak Load" tab is active.

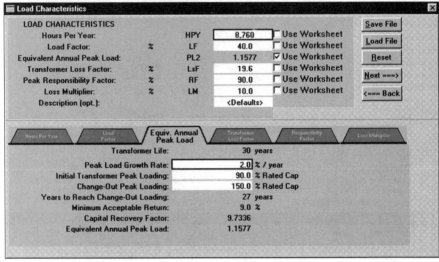

Figure 8: Equivalent Annual Peak Load Calculation Sheet in the Load Characteristics screen

In this worksheet as well as the others, cream colored cells may be edited if necessary.
To change a value, simply click into the desired cell and type the more accurate number.
Some calculation sheets allow you to enter information stored in files (e.g., load profile
files) by double clicking on cream colored cells. This is covered in the DTCEM Toolbox
Chapter (Chapter 4) of this manual.

In this case study, ABC Utility expects that the initial transformer loading of the desired
transformers will be 92% and the change-out peak loading will be 140%.

→   *Type 92 in the "Initial Transformer Peak Loading" box.*

→   *Type 140 in the "Change-Out Peak Loading" box.*

The Equivalent Annual Peak Load value changes to 1.1423 from a default value of 1.1577
after making these changes. These changes are reflected in both the top and bottom
portion of the Load Characteristics screen as shown in Figure 9.

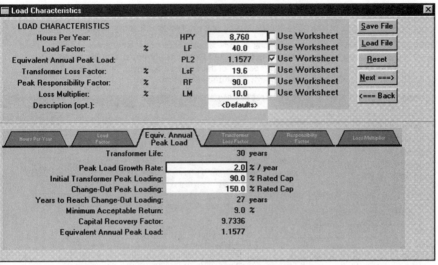

Figure 9: Changed Load Characteristics screen for ABC Utility

→   **Continue with the interview by clicking on the "Next" button.**

Step 6. Enter Cost Factors

The next step is to enter the costs of system capacity, energy, generation, and transmission/distribution avoided by using energy efficient transformers. The "Estimate Avoided Costs" screen is shown in Figure 10.

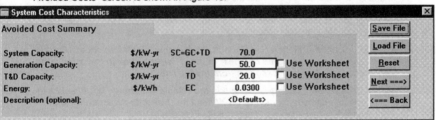

Figure 10: Cost Factors Screen

This screen includes default values for the avoided cost factors including:

* System Capacity (SC) (page 57)
* Generation Capacity (GC) (page 57)
* Transmission and Distribution (TD) (page 57)
* Energy (EC) (page 57)

These costs are used in the calculation of the value of the load and no-load losses. The costs may be entered in this screen in one of two ways. First, the costs may be directly entered by clicking in the corresponding cream colored cell and typing the correct values. Alternatively, the costs may be calculated by clicking on the corresponding "Use Worksheet" box to the right of the value and entering the information needed to calculate the avoided cost in the calculation sheet at the bottom of the screen.

For ABC Utility we will calculate the values for generation capacity (GC) and energy capacity (EC).

→   *Click in the "Use Worksheets" box next to the "Generation Capacity" cell.*

→   *Click in the "Use Worksheets" box next to the "Energy" cell.*

The Cost Factors screen expands to show a calculation sheet with three tabs in the lower portion of the screen as shown in Figure 11. Only tabs corresponding to clicked "Use Worksheets" boxes in the upper portion of the screen are active. In this case, the "Generation Capacity" and "Energy Cost" tabs are active.

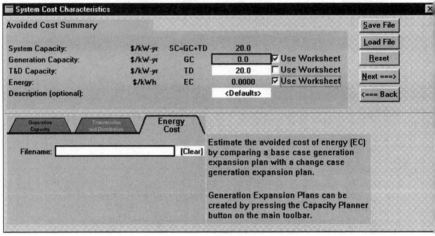

Figure 11: Expanded Cost Factors screen

In this worksheet as well as the others, cream colored cells may be edited if necessary. To change a value, simply click into the desired cell and input your utility's actual value. Some calculation sheets allow you to enter information stored in files (e.g., load profile files) by double clicking on cream colored cells. This is covered in the DTCEM Toolbox Chapter (Chapter 4) of this manual.

→   *Click on the "Generation Capacity" tab in the calculation sheet. Double click in cream colored box next to the "Filename" box in the calculation sheet. Select basecase.cap from the file lists.*

→   *Click on the "Energy Cost" tab in the calculation sheet. Double click in cream colored box next to the "Filename" box in the calculation sheet. Select "basecase.cap" from the file lists.*

The Generation Capacity value changes to 79.8 from a default value of 50.0 after making these changes and the Energy Capacity value changes to 0.0301 from a default value of 0.0300. These changes are reflected in both the top and bottom portion of the Cost Factors screen as shown in Figure 12.

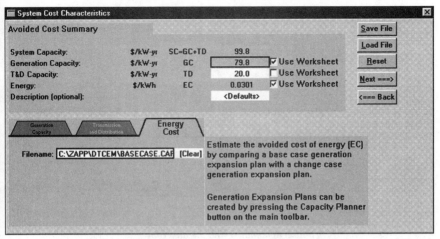

Figure 12: Changed Cost Factors Screen for ABC Utility

→   **Continue with this interview by clicking on the "Next" button.**

Step 7. Enter Transformer Bids

The next step is to enter the bids for the transformer made by the different suppliers. You should select **Yes** in the screen as shown in Figure 13 below:

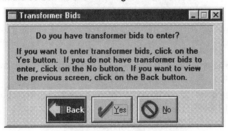

Figure 13: A Question About Bids screen

You should then select **Yes** in the screen as shown in Figure 14 below:

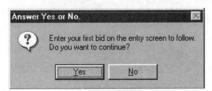

Figure 14: Answer Yes or No Screen

You will now see the "Enter a Transformer Bid" dialog box. In this screen you should enter the supplier's name, bid price, core losses, and winding losses.

→   **Enter "Manufacturer 1" in the Supplier Box.**

→   **Enter "390" in the Price Box.**

→   **Enter "95" in the Core Losses Box.**

→   **Enter "413" in the Winding Losses Box.**

The finished Transformer Dialog box should look like Figure 15.

Figure 15: Transformer Bid Dialog Box for Manufacturer 1

→   **Click on the OK button to continue with the interview.**

You should continue to answer **Yes** in the "Answer Yes or No" box (Figure 16) until you enter all 6 of the bids described at the top of this case study. Continue to enter the supplier's name, price, core losses, and winding losses as described above.

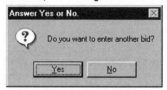

Figure 16: Answer Yes or No box

→   **When you have finished entering all six of the bids, click No in the "Answer Yes or No" Box.**

You have now completed the data entry portion of the case study. You should see an "All Done" box (Figure 17) indicating that you are finished.

Figure 17: All Done Dialog box

→   **Click "Finish" to end the interview.**

Your completed "Transformer Bid Evaluation Screen" pops up as shown in Figure 18 below.

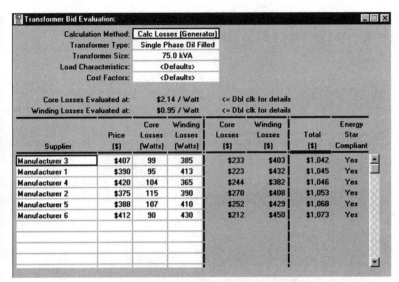

Figure 18: Completed Bid Evaluation Screen for ABC Utility

As you can see, the bids are ordered from the lowest total cost to the highest. Manufacturer 3 has the lowest total cost despite having the fourth highest bid price. The total cost is calculated by using the following formula:

Total Owning Cost = (Bid Price + Cost of Core Losses + Cost of Winding Losses)

Step 8. Analyze the Data

The next step is to view the comparative energy costs and benefits for the different bids in the Bid Evaluation screen.

→ *Click on the Quick Report icon (Figure 19) on the floating toolbar.*

You are first shown the "Report Control Panel" in which you may select the number of transformers you wish to purchase, the method by which you want to sort the transformers, the transformer bids you wish to analyze, and the transformer bid which you wish to serve as the "base case".

Figure 19: Quick Report icon

→ *Enter "500" in the Number of Units to Purchase box.*

→ *Click the "Lowest Total Owning Cost" option to sort the transformers.*

→ *Click the "Select All" button to select all of the bids to be analyzed.*

→ *Click on "Manufacturer 5" in the Bids Selected Box and then click on the "Select" button. (Assume that you would have purchased transformers from Manufacturer 5 before analyzing the total owning costs in DTCEM. This report will compare the costs and benefits of purchasing the transformers from the other manufacturers to the costs and benefits of purchasing Manufacturer 5's transformers.)*

Your finished "Report Control Panel" should look like Figure 20 below:

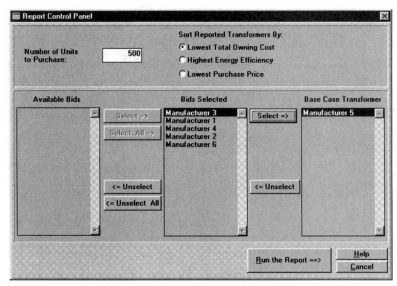

Figure 20: Report Control Panel for ABC Utility

→ **Click on the "Run the Report" button to run the report.**

The "Bid Evaluation Report" pops up as shown in Figure 21.

PER TRANSFORMER Supplier	Energy Star	Standard Efficiency at 50% Load	Total Owning Cost	Price	Price Diff.	Wattage Loss Watts	Energy Loss kWh	Energy Cost $/yr	Energy Savings $/yr	Simple Payback Years
Base=> Manufacturer 5	Yes	99.44%	$1,068	$388	$0	173	1,512	$45		
Manufacturer 3	Yes	99.48%	$1,042	$407	$19	161	1,407	$42	$3.16	6.0
Manufacturer 1	Yes	99.47%	$1,045	$390	$2	161	1,411	$42	$3.04	0.7
Manufacturer 4	Yes	99.48%	$1,046	$420	$32	162	1,423	$43	$2.69	11.9
Manufacturer 2	Yes	99.44%	$1,053	$375	-$13	177	1,554	$47	-$1.26	10.3
Manufacturer 6	Yes	99.48%	$1,073	$412	$24	159	1,391	$42	$3.64	6.6

Utility Info: ABC Utility, Indiana. Transformer: Single Phase Oil Filled, Size 75.0, Lifetime 30, Number 500. Date: 10/23/1996.

Figure 21: ABC Utility Bid Evaluation Report

This report screen is not maximized when it is displayed. It may be necessary to use the scroll bars on the right side and the bottom of the screen to scroll through the entire report.

The top part of the report contains information about the utility and the transformers. The bottom part of the report contains two tables, both of which analyze the energy costs, benefits, and emissions avoided. These values in the top row of the table are for the "base case" transformer from Manufacturer 6. The remaining rows display values for the other transformers relative to this base case.

If we compare the costs and benefits of Manufacturer 3's transformer to the basecase (Manufacturer 5) we see that despite costing $19 more per transformer ($9,500 more for 500 transformers) Manufacturer 3's transformer will save $3.16/year in energy ($1,581/year for 500 transformers)and will result in a 6.0 year payback period. Though the transformers from Manufacturer 3 will cost $9,500 more in initial capital costs, over the 30 year transformer lifetime, the transformers will save $47,430 ($1,757/year times 30 years) when compared to the transformers from Manufacturer 5.

It should be noted here that the simple payback may be a negative number. Simple payback is calculated by dividing the price difference by the energy savings/year. The table below outlines the potential scenarios and details what the simple payback in each of these cases means.

Scenario	Price Diff.	Energy Savings	Simple Payback
1	positive	positive	Payback is positive. The price for the transformer is higher and the energy savings are greater than the basecase. In the short term, the cost is higher but over the life of the transformer, the savings are greater.
2	negative	negative	Payback is positive. The price for the transformer is less and the energy savings are less than the basecase. In the short term, the cost is lower but over the life of the transformer the energy savings will be lower.
3	positive	negative	Payback is negative. The price for the transformer is higher and the energy savings are less than the basecase. This is the least desired scenario. The short term initial costs are higher and the savings over the life of the transformer are less.
4	negative	positive	Payback is negative. The price for the transformer is less and the energy savings are greater than the basecase. This is the best scenario. The short term costs are less and the savings over the life of the transformer are greater.

→   *Close this report by double clicking on the screen's upper left hand corner.*

It is recommended that you save the bid report file which you just entered to enable you to open the file later without reentering the data.

→   *Click on the save bid icon (Figure 22) on the toolbar and type the name of the file you wish to save.*

At this point you have entered and analyzed the transformer bids for ABC Utility. You may wish to edit the default data that was used in the calculation of the core and winding losses by examining the Capacity Planner, Trans/Dist Planner, and Load Profiles Planner. These features are described in the DTCEM Toolbox Chapter (Chapter 4) of this manual. If you have any additional questions about DTCEM or the ENERGY STAR Transformer Program, call the Energy Star Hotline at **1-888-STAR-YES** (toll-free).

Figure 22: Save bid icon

Chapter 2. Setting Up Defaults

DTCEM uses default values in the calculation of the costs and benefits of transformer bids. Options contained under the **Setup** menu allow you to make changes to these default values. These options include editing the **Utility Information**, the **Financial Data**, and the **Cost Data**. These options are described below.

Utility Information

To enter basic information about your utility, click on the setup icon (Figure 23) on the toolbar or select **Utility Information** from the **Setup** menu. The "Utility Information" dialog box pops up as shown below in Figure 24:

Figure 23: Setup icon

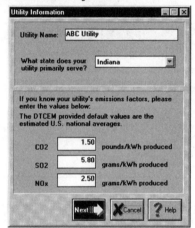

Figure 24: Utility Information Dialog Box

The following information should be entered in this screen:

*   Name of utility
*   State primarily served by utility
*   Emissions factors (for CO_2, SO_2, and NO_x)

Note: The default data for the emission factors are the national averages as calculated by the EPA.

Click on OK to save this information.

Financial Data

To enter basic financial data which will be used in the cost and benefit calculations, select **Financial Data** from the **Setup** menu. The "Enter Financial Factors" dialog box pops up as shown in Figure 25:

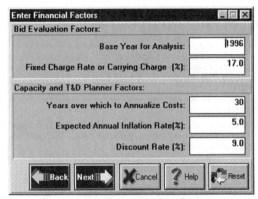

Figure 25: Financial Data Dialog Box

The following information should be entered in this screen:

- Base year for the analysis
- Annualization years (years for levelizing values)
- Annual inflation rate (%)
- Discount rate (%)
- Fixed charge rate (%)

Click on OK to save this information.

Cost Data

To edit the default costs associated with the capacity and fuels, select **Cost Data** from the **Setup** menu. You have two choices, **Capacity Costs** and **Fuel Costs**. Both are tables as shown in Figure 26 and Figure 27 below:

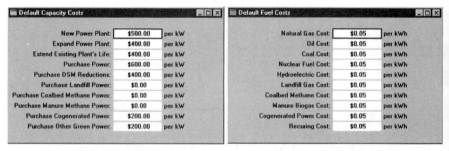

Figure 26: Default Capacity Costs Table   *Figure 27: Default Fuel Costs Table*

Any of the default costs in the cream colored boxes may be edited if necessary. Simply click into the box and type the correct value. These changes will be permanently saved. If you wish to reset *one* of the values to the defaults, click in the cream colored cell you wish to reset and click on the reset one icon (Figure 28) in the floating toolbar. If you wish to reset *all* of the values to the defaults, click on the reset all icon (Figure 29) on the floating toolbar.

Figure 28: Reset one icon

The natural gas, oil, and coal default fuel costs may be calculated using the heat rate calculator dialog. To access the Heat Rate Calculator Dialog (Figure 30), double click in the cream colored cell corresponding to either the natural gas, oil, or coal cost.

Figure 29: Reset all icon

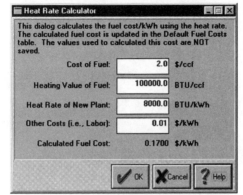

Figure 30: Heat Rate Calculator Dialog Box

The Heat Rate Calculator calculates the cost of fuel in $/kWh based on the following equation:

(CF / HV * HRNP) + OC

where:

CF	=	Cost of the fuel (mass)
HV	=	Heating Value of the fuel
HRNP	=	Heat Rate of the New Plant
OC	=	Other Costs (i.e., Labor)

Any of these values may be edited in this screen if necessary. Note that these values are NOT saved upon exiting the dialog. Only the calculated cost of fuel/kWh is saved and updated in the Default Fuel Costs Table.

Click on the OK button to save and exit this dialog.

You may exit the Default Fuel Cost and Default Capacity Cost tables by double clicking in the upper left hand corner or the screen.

Chapter 3. The Bid Evaluation Screen

The "Transformer Bid Evaluation" screen is where most of the information needed to analyze transformer costs and efficiencies needs to be entered. A blank bid evaluation screen can be opened by clicking on the New bid icon (Figure 31). This blank screen is shown below in Figure 32.

Figure 31: New bid icon

Transformer Bid Evaluation:[noname.bid]							
Calculation Method:	Calc Losses (Generator)						
Transformer Type:	Single Phase Oil Filled						
Transformer Size:	75.0 kVA						
Load Characteristics:	<Defaults>						
Cost Factors:	<Defaults>						
Core Losses Evaluated at:	$3.50 / Watt	<= Dbl clk for details					
Winding Losses Evaluated at:	$1.32 / Watt	<= Dbl clk for details					

Supplier	Price ($)	Core Losses (Watts)	Winding Losses (Watts)	Core Losses ($)	Winding Losses ($)	Total ($)

Figure 32: Blank bid evaluation screen

The following types of information are required in the "Transformer Bid Evaluation" screen:

- Calculation Method
- Transformer Size and Type
- Transformer Supplier and Price Information

The following sections describe this information and how to enter it into the DTCEM "Transformer Bid Evaluation" screen.

Calculation Method

Core and winding losses are utility specific variables that account for capitalized costs per rated Watt at no-load and full-load conditions. The cost of core losses are often referred to as the "A" factor and the cost of winding losses are often referred to as the "B" factor. These values may be derived using utility specific capital, fuel, generation, transmission, operation, and maintenance costs, along with customer demographics. When evaluating a transformer bid with DTCEM you should determine the calculation method which will calculate these variables. The choices you have in DTCEM are:

(1) Enter Losses Directly: use core and winding losses which have already been calculated.

(2) Calc Losses (Generator): calculate core and winding losses for a generating utility.

(3) Calc Losses (Disco): calculate core and winding losses for a distribution cooperative without generation.

You may select one of the options by double clicking on the cream colored cell next to the **Calculation Method** text at the top of the bid evaluation screen. The "Select the Calculation Method" dialog box is presented as shown in Figure 33 below:

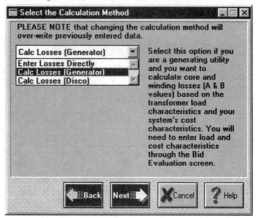

Figure 33: Calculation method dialog box

You should select one of the three options from the pick list. Click on OK to save and continue. Based on your selection, you may need to enter **Load Characteristics** and **Cost Factors** to calculate the A and B factors and the total transformer costs. This information is described in the sections below.

Entering Known A and B Values

The "Enter A/B Directly" selection should be made from the "Select a Calculation Method" dialog box if the costs of the core and winding losses (A and B) are known.

This selection allows you to input previously calculated A and B values without derivation of all of the parameters associated with the values. Consequently, upon making this selection, the boxes corresponding to the **Load Characteristics** and **Cost Factors** cells in the upper portion of the Bid Evaluation table change to display "<optional>" text and the boxes corresponding to the A and B values change to a cream color. These cream colored cells may be double clicked to open up the "Enter A and B Values" dialog box as shown in Figure 34 below:

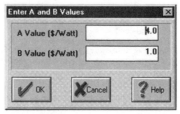

Figure 34: Enter A and B values dialog box

Enter the accurate calculated values for A and B to the nearest cent (e.g., $1.72/Watt) in the appropriate boxes. Click on OK to save and exit.

Using these defined A and B values, DTCEM calculates the total owning cost of each supplier's transformer using the following formula:

Total Owning Cost = (Bid Price + Cost of Core Losses + Cost of Winding Losses)

where:

Cost of Core Losses = (1.0 + Loss Multiplier) * A value * Core Losses

Cost of Winding Losses = (1.0 + Loss Multiplier) * B value * Winding Losses

---

Load Characteristics

If you are entering known A and B values you may optionally edit the default load factor and loss multiplier which are used in the calculation of the costs and benefits. To edit these load characteristics, double click on the **Load Characteristics** box in the **Bid Evaluation** screen. The "Enter Load Factor and Loss Multiplier" box pops up as shown in Figure 35 below.

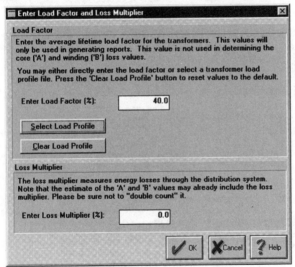

Figure 35: Load Factor and Loss Multiplier screen

Load Factor

The Load Factor (LF) is the average transformer load over a period of time. The load factor is calculated by dividing the energy consumed by the transformer's load by the maximum load available. A transformer that is loaded continuously with the maximum load therefore has a value of 1.0.

The LF may be entered directly in the "Enter Load Factor %" box or calculated from a **transformer load profile** by clicking on the "Select Load Profile" box.

Loss Multiplier

The Loss Multiplier (LM) is not calculated in the DTCEM program. This value may be entered in the "Enter Loss Multiplier (%)" box. The LM is based on the utility's overall transmission and distribution losses.

Cost Factors

If the Calculation Method is "Enter A/B Directly", the average annual energy cost per kWh may be optionally entered by double clicking on the cream colored cell corresponding to the **Cost Factors** cell in the top of the Bid Evaluation table. The "Enter the Estimated Value of Energy Saved" box pops up as shown in Figure 41 below:

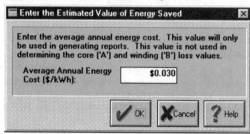

Figure 36: Estimated Value of Energy Saved

The average annual energy cost is used in generating the costs and benefits of the transformer bids. This value may be edited by clicking in the "Average Annual Energy Cost ($/kWh)" box and typing the correct value.

Calc A/B (Generator)

The "Calc A/B (Generator)" selection should be made from the "Select a Calculation Method" dialog box if you wish to calculate the core and winding losses (A and B values) for a generating utility. This selection requires that you enter **Load Characteristics** and **Cost Factors** information. DTCEM has provided defaults for this information; however, it is recommended that you double click on the cream colored cells corresponding to these factors and update the defaults using utility specific values.

Using these factors, DTCEM calculates the A and B values using the following formulae:

$$A = (SC + (EC * HPY)) / (FCR * 1000)$$

$$B = (((SC * RF) + (EC * LsF * HPY)) * PL2) / (FCR * 1000)$$

where:

SC	=	Avoided cost of system capacity
EC	=	Avoided cost of energy
HPY	=	Hours per year
FCR	=	Fixed charge rate
RF	=	Responsibility factor
LsF	=	Transformer Loss Factor
PL2	=	Equivalent Annual Peak Load

The total cost of the transformer for each supplier is then calculated using the following formula:

Total Owning Cost = (Bid Price + Cost of Core Losses + Cost of Winding Losses)

where:

Cost of Core Losses   = (1.0 + Loss Multiplier) * A value * Core Losses

Cost of Winding Losses = (1.0 + Loss Multiplier) * B value * Winding Losses

Load Characteristics

If you are calculating A and B factors for a generating utility or for a distribution cooperative it is recommended that you edit the default load characteristics which are used to derive these values. To edit the load characteristics, double click in the cream colored box corresponding to the **Load Characteristics** in the **Bid Evaluation** screen. The "Load Characteristics" box pops up as shown in Figure 37 below.

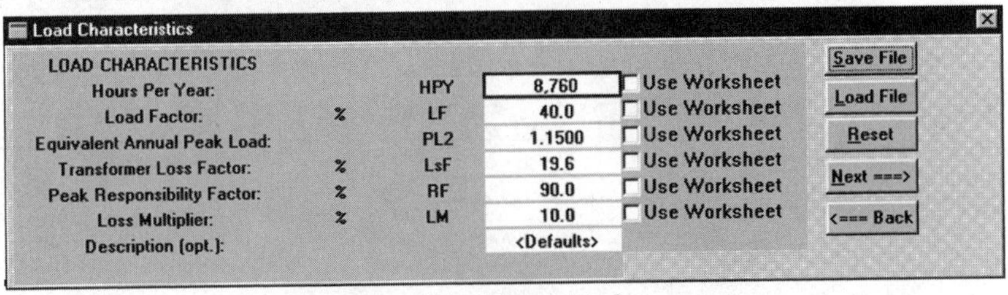

LOAD CHARACTERISTICS					
Hours Per Year:		HPY	8,760	Use Worksheet	Save File
Load Factor:	%	LF	40.0	Use Worksheet	Load File
Equivalent Annual Peak Load:		PL2	1.1500	Use Worksheet	Reset
Transformer Loss Factor:	%	LsF	19.6	Use Worksheet	Next ===>
Peak Responsibility Factor:	%	RF	90.0	Use Worksheet	
Loss Multiplier:	%	LM	10.0	Use Worksheet	<=== Back
Description (opt.):			<Defaults>		

Figure 37: Load Characteristics Box

The load characteristics box contains the transformer's operating load characteristics. There are six different parameters in this screen:

- Hours Per Year (HPY)
- Load Factor (LF)
- Equivalent Annual Peak Load (PL2)
- Transformer Loss Factor (LsF)
- Peak Responsibility Factor (RF)
- Loss Multiplier (LM)

In addition to these parameters, a description of the load characteristics may be entered in the **Description** box.

These factors may be entered directly in the cream colored box by clicking in the box and typing the correct value. The factors may also be calculated by using worksheets. To calculate a factor, click on the corresponding "Use Worksheets" box to the right of the cream colored box. A calculation sheet with six tabs is created in the lower portion of the screen as shown in Figure 38. Only tabs corresponding to clicked "Use Worksheets" boxes in the upper portion of the screen are active. While in a calculation sheet, any of the values in the cream colored boxes may be changed. Changed values are reflected in the upper portion of the "Load Characteristics" screen.

The following sections describe the calculation worksheets which correspond the to appropriate load characteristics.

Hours Per Year

Hours per year (HPY) is the number of hours that the transformer is expected to be energized is assumed to be 8,760 (24 hours per day times 365 days per year). However, there may be exceptions such as transformers connected to irrigation loads. If the hours per year (HPY) is not 8,760, you may enter the HPY in this table or click on the "Use Worksheet" box and enter the **hours per day** and **days per year**.

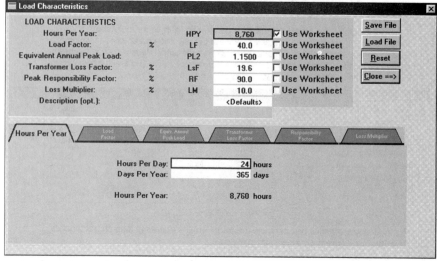

Figure 38: Expanded Load Characteristics screen

Load Factor

The Load Factor (LF) is the average transformer load over a period of time. The load factor is calculated by dividing the transformer's load by the capacity load available. A transformer that is loaded continuously with the same load therefore has a value of 1.0.

The LF may be calculated from a **transformer load profile** by double clicking on the box next to the "Select a Transformer Load Profile File" in the Load Factor calculation sheet.

Equivalent Annual Peak Load

The Equivalent Annual Peak Load (PL2) is the levelized annual peak load on the transformer that is equivalent to an initial peak load with an estimated load growth rate and a maximum allowable load before change out is required. The PL is calculated from the following factors which may be entered in the PL2 calculation sheet: **transformer life**, **peak load growth rate**, **initial transformer loading**, and **change-out loading**.

However, your utility may have historical data that should be used in place of the defaults.

Transformer Loss Factor

The Transformer Loss Factor (LsF) is a measure of the annual average load losses to the peak value of load losses on the distribution transformer. If the LsF is known, it may be entered in the corresponding box at the top of the Load Characteristics table. However, it may be calculated by clicking on the "Use Worksheets" box and entering the **annualized transformer load factor** (%) and/or the **transformer load profile file**.

Peak Responsibility Factor

The Peak Responsibility Factor (RF) is a measure of the diversity of the load on the transformer. If the RF is not known, it may be calculated by clicking on the "Use Worksheets" box. The calculation sheet for RF gives two options for calculating this factor. If both the **transformer's load at the time of the system's load** and the **transformer's peak load** are known, these values may be entered in the appropriate field. If these factors are unknown, the **transformer's load profile file** and the **system's load profile file** for the same period (daily, monthly, or hourly) may be used.

Loss Multiplier

The Loss Multiplier (LM) is not calculated in the DTCEM program. This value may be entered directly in the top portion of the Load Characteristics screen or in the LM calculation sheet by clicking on the corresponding "Use Worksheets" box. The LM is based on the utility's overall transmission and distribution losses. It shows the impact of distribution and transmission losses on generation equipment.

Saving, Retrieving, and Resetting Load Characteristics

Load Characteristics may be saved by clicking on the "Save" button and may be retrieved by clicking on the "Retrieve" button in the upper right hand corner of the Load Characteristics screen. These values may be reset to the default values by clicking on the "Reset to Defaults" button.

Cost Factors

The capacity and energy costs from transformer inefficiencies are used to calculate the A and B values. If the Calculation Method is "Calc A/B (Generator)", these avoided costs must therefore be entered to accurately derive these values.

The following types of avoided costs may be entered and edited:

- Avoided Cost of System Capacity (SC)
- Avoided Cost of Generation Capacity (GC)
- Avoided Cost of T&D Capacity (TD)
- Avoided Cost of Energy (EC)

To edit these Cost Factors, double click on the cream colored cell corresponding to the **Cost Factors** cell in the top of the Bid Evaluation table. The "Estimate Avoided Costs" screen pops up as shown in Figure 39 below.

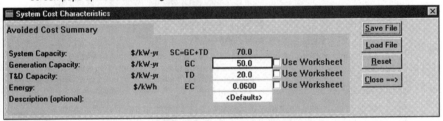

Figure 39: Cost Factors screen for the TOC Method

The values for GC, TD, and EC may be entered directly by clicking in the corresponding box and typing the accurate value. The factors may also be calculated by using worksheets. To access a calculation worksheet for either GC, TD, or EC, click in the corresponding "Use Worksheet" check box to the right of the value. The "Estimate Avoided Costs" screen expands exposing a lower section with tabulated tables corresponding to the GC, TD, and EC values in the Avoided Cost Summary table as shown in Figure 40. Only tabs corresponding to clicked "Use Worksheets" boxes in the upper portion of the screen are active.

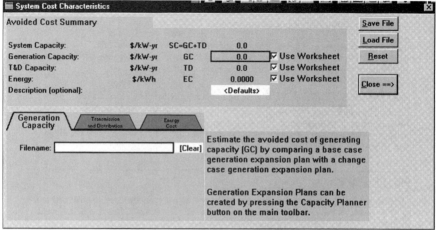

Figure 40: Estimate Avoided Costs Using Worksheets screen

The following sections describe the calculation worksheets corresponding to the characteristics contained within this "Estimate Avoided Costs" screen.

Avoided Cost of System Capacity

The Avoided Cost of System Capacity (SC) is the levelized avoided (incremental) cost of generation, transmission, and primary distribution capacity required to supply the next kW of load to the distribution transformer coincident with the peak load. SC is calculated by adding the Avoided Cost of Generation Capacity (GC) and the Avoided Cost of T&D Capacity (TD). A value for SC may not be entered directly.

Avoided Cost of Generation Capacity

The Avoided Cost of Generation Capacity (GC) is the incremental cost of adding generating capacity to the utility's system. This value is typically shown on a $/kW basis. There may be multiple scenarios for the installation of new capacity, and the avoided cost typically represents the costs (or the savings) from using the least costly scenario.

If the GC value is not known, it may be calculated by clicking in the "Use Worksheets" box. Base case and change case files should be selected which will then be analyzed to calculate GC (and EC, see below). These files may be entered and saved as discussed on page 39 in the **Generation & Energy (GC & EC)** section. The difference between the system capacity and total cost under each "case" or scenario will determine the avoided cost of capacity.

Avoided Cost of T&D Capacity

The Avoided Cost of T&D Capacity (TD) is the incremental cost of adding transmission and distribution capacity to the utility's system. This value is typically shown on a $/kW basis. There may be multiple scenarios for the installation of new T&D capacity, and the avoided cost typically represents the costs (or the savings) from using the least costly scenario.

If the TD value is not known, it may be calculated by clicking in the "Use Worksheets" box. A Transmission and Distribution file should be selected which will then be analyzed to calculate TD. This file may be entered and saved as discussed on page 44 in **Transmission & Distribution (TD)** section.

Avoided Cost of Energy

The Avoided Cost of Energy (EC) is the levelized avoided (incremental) cost for the next kWh produced by the utility's generating units. If the EC value is not known, it may be calculated by clicking in the "Use Worksheets" box. Base case and change case files should be selected which will then be analyzed to calculate EC. These files may be entered and saved as discussed on page 39 in the **Generation & Energy (GC & EC)** section. The difference in fuel costs between the two "cases" or scenarios will determine the EC.

Saving, Retrieving, and Resetting Cost Factors

Cost Factors may be saved by clicking on the "Save" button and may be retrieved by clicking on the "Retrieve" button in the upper right hand corner of the Cost Factors screen. These values may be reset to the default values by clicking on the "Reset to Defaults" button.

Calc A/B (Disco)

The "Calc A/B (Disco)" selection should be made from the "Select a Calculation Method" dialog box if you are a distribution cooperative without generating capability and wish to calculate the A and B Values. This selection requires that you enter **Load Characteristics** and **Cost Factors** information. DTCEM has provided defaults for this information however it is recommended that you double click on the cream colored cells corresponding to these factors and update the defaults using utility specific values.

Using these factors, DTCEM calculates the A and B values using the following formulae:

$$A = ((12.0 * DC) + (HPY * EC)) / (FCR * 1000)$$

$$B = ((PL2 * RF^2 * 12.0 * DC) + (PL2 * HPY * LsF * EC)) / (FCR * 1000)$$

where:

DC	=	Demand charge
EC	=	Energy charge
HPY	=	Hours per year
FCR	=	Fixed charge rate
RF	=	Responsibility factor
LsF	=	Transformer Loss Factor
PL2	=	Equivalent Annual Peak Load

The total cost of the transformer for each supplier is then calculated using the following formula:

Total Owning Cost = (Bid Price + Cost of Core Losses + Cost of Winding Losses)

where:

Cost of Core Losses   = (1.0 + Loss Multiplier) * A value * Core Losses

Cost of Winding Losses = (1.0 + Loss Multiplier) * B value * Winding Losses

Load Characteristics

If you are calculating A and B factors for a distribution cooperative ("Calc A/B (Disco)"), it is recommended that you edit the default load characteristics which are used to derive these values. To edit the load characteristics, double click on the cream colored cell corresponding to **Load Characteristics** in the **Bid Evaluation** screen. The **Load Characteristics** box and the data it requires are described in the Load Characteristics section under the "Calc A/B (Generator)" section on page 27 above.

Cost Factors

If the Calculation Method is "Calc A/B (Disco)", the demand charge and energy charge must be entered to derive the A and B factors. Distribution cooperatives that do not have generating capability will not experience avoided costs of generation and transmission/distribution by using energy efficient transformers. These costs are therefore not accounted for in the calculation of the total cost for a transformer using the life cycle cost methodology. Instead, the demand and energy charges should be entered by double clicking on the cream colored cell corresponding to the **Cost Factors** cell in the top of the Bid Evaluation table. The "Cost Factors" box pops up as shown in Figure 41 below:

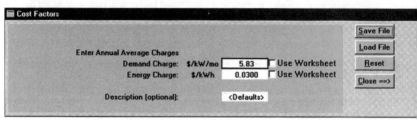

Figure 41: Cost Factors screen for the Calc A/B (Disco) Method

The values for the demand charge and energy charge may be entered directly by clicking in the corresponding box and typing the accurate value. The factors may also be calculated by using worksheets. To access a calculation worksheet for either the demand charge or the energy charge, click in the corresponding "Use Worksheet" check box to the right of the value. The "Cost Factors" screen expands exposing a lower section with tabulated tables corresponding to the demand charge and energy charge as shown in Figure 42. Only tabs corresponding to clicked "Use Worksheets" boxes in the upper portion of the screen are active.

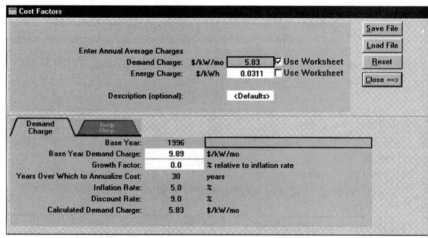

Figure 42: Expanded Cost Factors screen for the Calc A/B Disco Method

Demand Charge

The demand charge is the per kW of electricity used. These charges typically vary by time of year and peak utility season. The worksheet for this charge averages the demand charge over the transformer life using the inflation rate and the growth factor. The charge for the base year and the growth factor may be edited in the cream colored boxes in this screen. The other factors: base year, inflation rate, and transformer life may be edited in the **Financial Data** screen. The average demand charge is displayed in the top portion of the screen.

Energy Charge

The energy charge is the per kWh of electricity used. These charges vary by both time of year, time of day, and amount used (block rates). The worksheet for this charge averages the energy charge over the transformer life using the inflation rate and the growth factor. The charge for the base year and the growth factor may be edited in the cream colored boxes in this screen. The other factors: base year, inflation rate, and transformer life may be edited in the **Financial Data** screen. The average energy charge is displayed in the top portion of the screen.

Saving, Retrieving, and Resetting Cost Factors

Cost Factors may be saved by clicking on the "Save" button and may be retrieved by clicking on the "Retrieve" button in the upper right hand corner of the Load Characteristics screen. These values may be reset to the default values by clicking on the "Reset to Defaults" button.

Transformer Size and Type

Regardless of the calculation method, the type, size, and lifetime of the transformer for which the bids are being evaluated must be entered. This information may be entered by double clicking on either the **Transformer Size** or **Transformer Type** cream colored box at the top of the Bid Evaluation table. The "Select Transformer Type and Size" dialog box pops up as shown in Figure 43 below.

Figure 43: Transformer Type and Size dialog box

Three drop down lists are presented in this dialog. The top box shows the type of transformer, the middle box shows the size in kVA, and the bottom box shows the transformer life. Any of the three parameters may be changed by clicking on the arrow to the right of the edit box. When you are finished, click on the OK button to exit. The changed size and/or type will be reflected in the bid evaluation screen.

Transformer Supplier and Price Information

The Transformer Supplier and Price Information comprises the information in the bottom portion of the Bid Evaluation screen. Specific information for each transformer is needed to perform the actual bid evaluation including:

- Supplier's name
- Price
- Core Losses (Watts)
- Winding Losses (Watts)

These characteristics may be entered by double clicking in a cream colored cell in one of the rows of the Bid Evaluation Table. The "Enter a Transformer Bid" dialog pops up as shown in Figure 44 below:

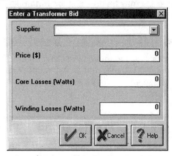

Figure 44: Transformer Bid Dialog

The information entered in this dialog is displayed in a row in the Bid Evaluation table. The loss figures are multiplied by their appropriate incremental cost values to estimate load loss and no-load loss costs. These values are used to calculate the first year losses and the lifetime losses. As the bids are added to the Bid Evaluation table, they are ranked and listed based on the lowest total owning cost (TOC).

Bid Evaluation Floating Toolbar

The floating toolbar (Figure 45) contains several options which allow you to edit the information in the bottom part of the Bid Evaluation table. These options are described below:

Clear All

To clear all of the bids from the lower part of the Bid Evaluation table click on the *Clear All icon* on the floating toolbar.

Delete

To delete one of the bids from the bottom part of the Bid Evaluation table click in the row containing the bid to be deleted and then click on the *Delete icon* on the floating toolbar.

Edit

To edit one of the bids in the bottom part of the Bid Evaluation table, click in the row containing the bid to be edited and then click on the *Edit icon* on the floating toolbar.

Figure 45: Bid Evaluation Floating toolbar

Quick Report

To view a report showing the comparative energy costs and benefits for the different bids in the Bid Evaluation click on the *Quick Reports icon* on the floating toolbar. *Note: At least two bids should be entered in the lower portion of the Bid Evaluation table before running this report.*

You will first be shown the "Reports Control Panel" (Figure 46) in which you may select the number of transformers you wish to purchase, the method by which you want to sort the transformers, the transformer bids you wish to analyze, and the transformer bid which you wish to serve as the "base case."

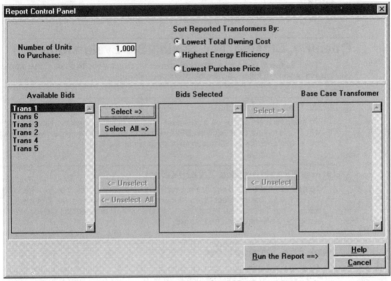

Figure 46: Reports Control Panel

When you have selected all of the information needed you may view the report by clicking on the "Run the Report" button in the Control Panel.

The top part of the report contains information about the utility and the transformers. The bottom part of the report contains two tables, both of which analyze the energy costs, benefits, and emissions avoided. These values in the top row of the table are for the "base case" transformers. The remaining rows display values for the other transformers relative to this base case.

You may print a hard copy of this report by clicking on the print icon on the toolbar. To close this report, double click on the upper left hand corner.

Summary Report

The summary report is similar to the Quick Report in that it shows the comparative energy costs and benefits for the different transformer bids in the Bid Evaluation table. This report, however, may be easily edited in any word processor and printed. To run the summary report, click on the *Summary Report icon* in the floating toolbar.

Like the Quick Report, you are first shown the Reports Control Panel (Figure 46) in which you may select the number of transformers you wish to purchase, the method by which you want to sort the transformers, the transformer bids you wish to analyze, and the transformer bid which you wish to serve as the "base case."

When you have selected all of the information needed you may view the report by clicking on the "Run the Report" button in the Control Panel.

The Summary Report is displayed in DTCEM's word processor. This file is called "reports.rtf." It is in rich text format which allows it to be opened in any word processor. You may edit this file or clean it up as necessary. This report contains a series of tables showing general information, technical parameters, and financial parameters regarding the transformer bids.

DTCEM's word processor may be closed by double clicking on the upper left hand corner

Close

To close the bid evaluation table, click on the *Close icon* in the floating toolbar.

Opening, Closing, and Saving Bid Evaluations

The **File** menu contains a variety of options which allow you to open, close, and save bid evaluations. These features are described below.

Figure 47: New bid icon

New Bid Evaluation

To evaluate a set of bids on a particular transformer rating, click the New bid icon (Figure 47) on the toolbar or select **New Bid Evaluation** from the **File** menu. You may have more than one new bid evaluation window open at one time.

Figure 48: Open bid icon

Open Existing Bid Evaluation

To open an existing bid evaluation which has been saved, click the Open Bid icon (Figure 48) on the toolbar or select **Open Existing Bid Evaluation** from the **File** menu. Type the name of the file you wish to open in the Filename box, or select the file you want to open using the drive, directory, and file lists.

Figure 49: Save bid icon

Save Bid Evaluation

To save a bid evaluation, click the Save Bid icon (Figure 49) on the toolbar or select **Save Bid Evaluation** from the **File** menu. When using Save, the copy you have been working on replaces the saved copy on disk. If you have not saved the session before, DTCEM will prompt you to name it.

Save Bid Evaluation as...

To save a new bid evaluation or to save the current bid evaluation with a new name or in a different directory, select **Save Bid Evaluation as...** from the **File** menu. Type the name of the file or select the file you want to replace. Choose OK to save the file.

Figure 50: Print icon

Print

To print a hard copy of the bid evaluation, select **Print** from the **File** menu or click on the Print icon (Figure 50) on the toolbar. This report is generated in the DTCEM word processor. Upon generation you may edit the report as needed. You may save this report by selecting Save from the DTCEM word processor File menu. To exit this word processor, double click in the upper left hand corner of the screen.

Exit

To exit the DTCEM Program, click on the Exit icon (Figure 51) on the toolbar or select Exit from the File menu. *Note: Any work which has not been saved prior to exiting will be erased. Be sure to save any important work before exiting.*

Figure 51: Exit icon

Chapter 4. DTCEM Toolbox

The DTCEM Toolbox consists of three options which allow you to edit some of the key parameters used in calculating the total owning cost (TOC) of a transformer. These toolbox options are contained under the **Tools** menu and allow you to enter information about the avoided cost of energy (EC), avoided cost of generation capacity (GC), and avoided cost of transmission and distribution capacity (TD).

Capacity Planner (GC & EC)

The capacity planner tool allows you to estimate the avoided costs of generation capacity (GC) and energy capacity (EC). The approach taken by the capacity planner is to develop a base case and change case scenario for providing capacity and energy. The base case represents the costs of providing a certain capacity. The change case is generated by decrementing (lowering) the capacity requirements by some amount (e.g., 10%) and re-estimating the cost of providing the necessary capacity. Because in most cases capacity expansion plans can be delayed by some number of years, the present value costs between the two cases represents the costs avoided by decrementing the system's capacity.

Figure 52: Capacity Planner icon

To generate a capacity expansion plan you need to select the period of analysis (e.g. 1996 - 2030), the system's peak load requirements (MW), and a means for achieving the capacity requirements (e.g., new power plants, purchasing power, DSM). You can enter up to 30 capacity source that can go "on-line" or "off-line" at any time you wish during the period of analysis. (If your selected capacity options do not meet your specified target requirement, DTCEM will indicate this with a red typeface in the "Sys Cap (MW)" column of the worksheet.) Each capacity source has associated with it a capital cost and operating cost which can be modified as necessary. DTCEM will use these capital and operating costs to estimate the total capital and operating cost of this "base case" expansion plan scenario.

After defining this base case expansion plan, the next step is to "decrement" it by some amount (e.g., 10%) and then re-evaluate the requirements for meeting this lower capacity requirement. DTCEM will perform this evaluation automatically by determining the maximum amount of time that each capacity source can be "deferred." Deferring one or more of the capacity sources by one or more years will lead to a savings in the net present value of the capital costs and operating costs. By dividing the capital cost savings by the amount of capacity decremented results in an estimate of the avoided cost of capacity in $/MW. Similarly, dividing the operating cost savings by the amount of energy generation avoided results in an estimate of the avoided cost of energy in $/kWh.

To bring up a worksheet to calculate the avoided cost of generation capacity and the avoided cost of energy, click on the Capacity Planner icon (Figure 52) on the toolbar or choose **Generation & Energy (GC & EC)** from the **Tools** menu. The "Capacity Plan" table pops up as shown in Figure 53 below.

	Peak Load (MW)	Load & Rsrv (MW)	Sys Cap (MW)	·	·	·	·	·	·	·	·
	Base	Base	Base								
1996	500	575	0								
1997	510	586	0								
1998	520	598	0								
1999	531	610	0								
2000	541	622	0								
2001	552	635	0								
2002	563	648	0								
2003	574	660	0								
2004	586	674	0								
2005	598	687	0								
2006	609	701	0								
2007	622	715	0								
2008	634	729	0								
2009	647	744	0								
2010	660	759	0								
2011	673	774	0								

Figure 53: Capacity Planner Table

In addition, a floating Capacity Planner "Control Panel" is displayed on top of the table. This control panel contains several icons which allow you to enter and edit the information detailed in the Capacity Plan. In addition, this Control Panel shows the output created from the information entered. The icons and the output table are described at the end of this section under **Capacity Planner Control Panel** on page 43.

A series of steps should be followed to ensure that all of the necessary information is added to this table. These steps are detailed below:

Set the Years of Analysis

The first thing you should do is set the years for which you are planning the capacity by clicking on any one of the years in the first column. The "Set the Years of Analysis" dialog box (Figure 54) pops up in which you should enter the start and end years for the period of analysis. Click on OK to save and continue.

Figure 54: Set Years Dialog Box

Enter the Capacity Needed

The next step is to enter the capacity needed for the years of analysis. To bring up the
Enter the Capacity Needed Dialog Box (Figure 55), double click in the **Peak Load**
column.

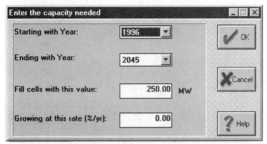

Figure 55: Enter the Capacity Needed Dialog Box

Use the picklists to enter a starting year and ending year, and enter the capacity in MW.
For smaller units (< 1 MW) use decimals to enter the capacity (e.g., 200 kW is 0.2 MW).
Enter the peak load growth rate, if applicable. Click OK to accept this information.

Set the Reserve Margin

The next step is to enter the load and reserve margin expected for this capacity. To bring
up the "Set the Reserve Margin" dialog box (Figure 56), double click in the Load and
Reserve column.

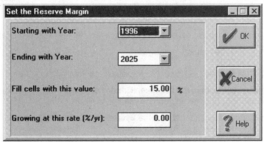

Figure 56: Set Reserve Margin Dialog Box

This dialog box allows you to select the reserve margin capacity needed for a given period
of analysis. Use the picklists to enter a starting year and ending year. Enter the percent
reserve margin in the box labeled "Fill Cells With This Value". The percent entered will be
added to the capacity needed in the **Peak Load** column and displayed in the
corresponding cells in the **Load and Reserve** column. Enter the peak load growth rate, if
applicable. Click on OK to save and continue.

Create Capacity

The next step is to add the capacity to be supplied. To bring up the "Create Capacity"
dialog box (Figure 57), double click on a cell in the fourth column or higher of the Capacity
Planner window.

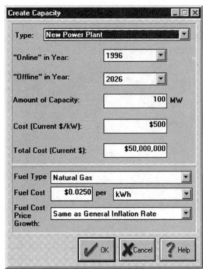

Figure 57: Create Capacity Dialog Box

In the top part of the screen, select or enter the following capacity information:

- Type of power or reduction in power (new power plant, power agreement purchase, etc.)
- Years the power (e.g., power plant) will be on-line
- Amount of effective capacity (MW)
- Cost per kW of this new capacity

In the bottom of the dialog box, select or enter the following fuel information:

- Fuel Type (natural gas, oil, etc.)
- Fuel Cost
- Fuel Cost Price Growth (at the defined general inflation rate)

Click on OK to save the changes. Created capacities will be displayed in the rows corresponding to the appropriate years.

Continue to add capacity until the **System Capacity** column values are greater than or equal to the **Load and Reserve** column (*Note: System capacity values that are greater than or equal to the Load and Reserve values are displayed in the System Capacity column in black text. Values less than the Load and Reserve values are displayed in red text.*).

Set the Decrement

Figure 58: Decrement icon

The next step is to enter the decrement value for the generation of the change case capacity plan. Click on the decrement icon (Figure 58) in the **Capacity Planner Control Panel** to bring up the "Enter the Amount in MW to Decrement Capacity By" dialog box (Figure 59).

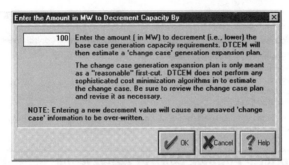

Figure 59: Decrement Dialog Box

You should enter the amount in MW to decrement (i.e., lower) the base case generation capacity requirements. DTCEM then calculates a 'change case' generation expansion plan.

Enter the decrement value in the white edit box. Click on OK to save and continue.

Save the Capacity Plan

Figure 60: Save icon

The final step is to save the capacity plan as a *.cap file. This file may be used later in the Cost Factors table to calculate GC and EC. Click on the save icon (Figure 60) in the Capacity Planner Control Panel or click F6. Saved *.cap files may be retrieved in later sessions by clicking on the Open icon (Figure 61) in the Capacity Planner Control Panel or clicking F5.

Figure 61: Open icon

Capacity Planner Control Panel

The Capacity Planner "Control Panel" (Figure 62) contains a table outlining the results of adding capacity to the Capacity Planner Table and a series of icons which may be used to access additional features of the Capacity Planner.

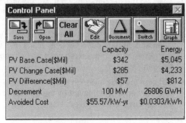

Figure 62: Capacity Planner Control Panel

This control panel is displayed over the top of the Capacity Planner table. You may move the control panel but you may not close it.

Capacity Planner Control Panel Table

The following values are contained within the Capacity Planner Control Panel table:

- Present value (PV) Base Case
- Present value (PV) Change
- Present value (PV) Difference
- Decrement
- Avoided Cost

These values are displayed for the capacity (how much is costs to purchase the capacity) and energy (how much it costs to operate the capacity).

Capacity Planner Control Panel Icons

A set of icons are presented at the top of the Capacity Planner Control Panel allowing you to operate some additional controls in the Capacity Planner. The save, open, and decrement icons have already been described in previous sections. The following additional features are accessible through these icons.

Save icon:	Click on the save icon to save a capacity plan (*.cap file).	
Open icon:	Click on the open icon to open a capacity plan (*.cap file).	
Clear All icon:	Click on the clear all icon to clear all of the added capacities.	
Edit icon:	Click on the edit icon to edit one of the added capacities.	
Decrement icon:	Click on the decrement icon to set the value for the generation of the change case capacity plan.	
Switch icon:	Click on the switch icon to switch between viewing the base and change case capacity plans (Be sure to set the decrement first or the base case and change case will be identical).	
Graph icon:	Click on the graph icon to graph the capacity required and the capacity achieved by the capacity plan throughout the years of analysis.	

Transmission/Distribution Planner (TD)

Figure 63:
Trans/Dist Planner
icon

The purpose of the Transmission/Distribution Planner tool is to estimate the avoided cost of transmission and distribution capacity. Unlike the approach taken by the Capacity Planner tool (described above and which relies on projected future costs), the Transmission/Distribution Planner relies on historical information to estimate the incremental cost of providing transmission and distribution capacity. The reason for the difference in approach is the difficulty in projecting future costs for transmission and distribution capacity. The historical data necessary for the approach used by the Transmission/Distribution Planner is generally available at most utilities. However, the applicability of these historical costs as the basis for estimating future incremental transmission and distribution capacity costs should be carefully considered.

The Transmission/Distribution Planner estimates the incremental costs of transmission and distribution using the slope coefficient from a linear regression (i.e. least-squares fit) of cumulative capacity investments on cumulative load growth. The relationship is:

$TD = **intercept** + **slope** * (System Capacity in MW)

The slope of this line can be interpreted as the marginal cost of providing an additional increment of transmission and distribution capacity. In estimating the regression lines the cumulative capacity investments should be suitably adjusted for inflation (e.g., expressed in 1996 dollars).

The Transmission/Distribution Planner allows you to enter capacity and investments over a period from 1970 to 1996. Values need only be entered where data are available. The Transmission/Distribution Planner adjusts the investment dollars for inflation. The inflation values used may be modified as necessary. After the data is entered, the regression can be run and the estimated relationship graphed. Details for using each of the Transmission/Distribution Planner tools are described below.

To start the Transmission/Distribution Planner, click on the Trans/Dist Planner icon (Figure 63) on the toolbar or choose **Transmission & Distribution (TD)** from the **Tools** menu. The "Transmission/Distribution Plan" table pops up as shown in Figure 64 below.

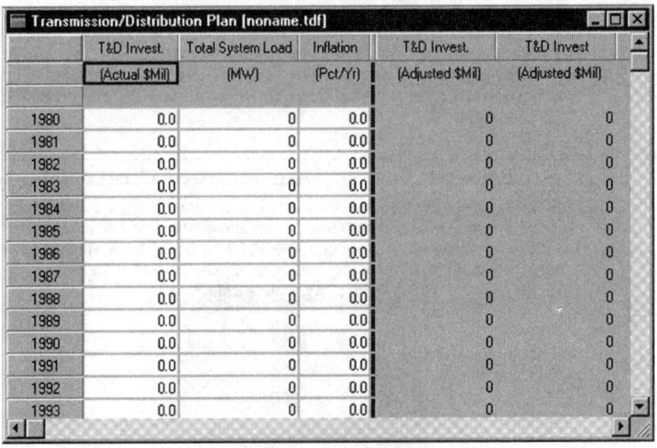

	T&D Invest.	Total System Load	Inflation	T&D Invest.	T&D Invest
	(Actual $Mil)	(MW)	(Pct/Yr)	(Adjusted $Mil)	(Adjusted $Mil)
1980	0.0	0	0.0	0	0
1981	0.0	0	0.0	0	0
1982	0.0	0	0.0	0	0
1983	0.0	0	0.0	0	0
1984	0.0	0	0.0	0	0
1985	0.0	0	0.0	0	0
1986	0.0	0	0.0	0	0
1987	0.0	0	0.0	0	0
1988	0.0	0	0.0	0	0
1989	0.0	0	0.0	0	0
1990	0.0	0	0.0	0	0
1991	0.0	0	0.0	0	0
1992	0.0	0	0.0	0	0
1993	0.0	0	0.0	0	0

Figure 64: Transmission/Distribution Plan

In addition, a floating Transmission/Distribution Planner Control Panel is displayed on top of the table. This control panel contains several icons which allow you to enter and edit the information detailed in the Transmission/Distribution Plan. In addition, this Control Panel shows the output created from the information entered. The icons and the output table are described at the end of this section under **Transmission/Distribution Planner Control Panel** on page 47.

A series of steps should be followed to ensure that all of the necessary information is added to this table. These steps are detailed below:

Set the Years of Analysis

The years of analysis are currently set to 1970 to 1996 and cannot be changed. A subsequent version of DTCEM will allow you to modify the analysis years. Please note that you do not need to enter data for every year. You may enter data for the years you have data available.

Enter Transmission/Distribution Investment

The next step is to enter the transmission and distribution investment in *actual* dollars. These values directly entered into the appropriate cream colored cells. Dollar values should be entered in millions of dollars (e.g., $10 million dollars should be entered as 10).

Enter Total System Load

The next step is to enter the total *cumulative* system load in MW. These values can be directly entered into the appropriate cream colored cell. DTCEM will not let you enter a cumulative system load that is less than the previous year's cumulative system load.

Enter Inflation Rate

The next step is to review the inflation rate in percent per year for the corresponding years of analysis. DTCEM provides you with default inflation rate values based on the U.S. Federal Reserve Board GDP deflator. These values can be edited as necessary.

Figure 65: Save icon

Save the Transmission/Distribution Plan

The final step is to save the transmission/distribution plan as a *.tdf file. This file may be used later in the Cost Factors table to calculate TD. Click on the save icon (Figure 65) in the Trans/Dist Planner Control Panel or click F6. Saved *.tdf files may be retrieved in later sessions by clicking on the Open icon (Figure 66) in the Trans/Dist Planner Control Panel or clicking F5.

Figure 66: Open icon

Transmission/Distribution Planner Control Panel

The Transmission/Distribution Planner "Control Panel" (Figure 67) contains a table outlining the results of adding transmission and distribution capacity to the Transmission/Distribution Planner Table and a series of icons which may be used to access additional features of the Transmission/Distribution Planner.

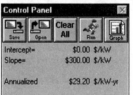

Figure 67: Transmission/Distribution Planner Control Panel

This control panel is displayed over the top of the Transmission/Distribution Planner table. You may move the control panel but you may not close it.

Transmission/Distribution Planner Control Panel Table

The following values are contained within the Transmission/Distribution Planner Control Panel table:

- Intercept
- Slope

- Annualized

The intercept is the no load transmission and distribution cost per kW. The slope is equal to the marginal, or incremental, cost of installing T&D capacity ($/kW). The annualized value is the annualized cost of transmission and distribution per kW.

Transmission/Distribution Planner Control Panel Icons

A set of icons are presented at the top of the Transmission/Distribution Planner Control Panel allowing you to operate some additional controls in the Transmission/Distribution Planner. These icons and their corresponding features are described below:

Save icon: Click on the save icon to save a trans/dist plan (*.tdf file).

Open icon: Click on the open icon to open a trans/dist plan (*.tdf file).

Clear All icon: Click on the clear all icon to clear all of the added transmission/distribution capacities.

Run icon: Click on the edit icon to run the regression analysis and see additional details for the regression analysis.

Graph icon: Click on the graph icon to graph the regression analysis. The graph displays the least-squares regression line along with the actual data points.

Load Profile Builder

The Load Profile Builder creates load profiles which are used in the calculation of the Load Factor, Transformer Loss Factor, and Peak Responsibility Factor. To create a load profile, click on the load profile builder icon (Figure 68) on the toolbar or select **Load Profile Builder** from the **Tools** menu. The Load Profile dialog box is displayed as shown in Figure 69 below.

Figure 68: Load profile builder icon

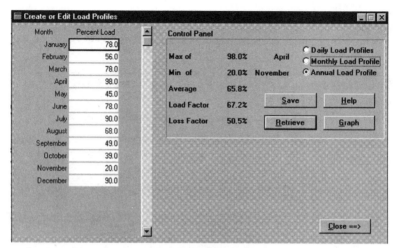

Figure 69: Load Profile Dialog Box

A daily, monthly, or yearly load profile may be entered by clicking on the appropriate button in the upper right hand corner of this box. The percent load should be entered in the corresponding cream colored boxes.

When you have completed entering the load profile values, you may save, edit, or graph the load profile by clicking on the appropriate buttons. Saved load profiles may be retrieved in the **Load Characteristics** screen when entering values in the calculation sheets for the Load Factor, Transformer Loss Factor, and Peak Responsibility Factor. Load profile graphs (Figure 70) may be edited using the **Graph** menu which is created upon graphing a load profile.

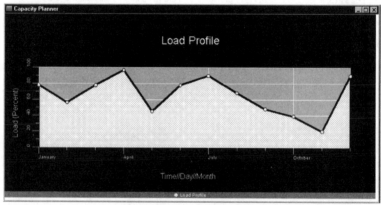

Figure 70: Sample Load Profile Graph

To close a graph, double click on the upper left hand corner of the graph's window. To exit the Load Profile Builder, click on the Close button.

Chapter 5. Help and Other Features

DTCEM contains help and other features which may assist you when using the program. These features are described below.

Help

The **Help** menu provides information about DTCEM's features and options through DTCEM's on-line Help system.

Contents

Select **Contents** from the **Help** menu or press F1 to see a list of the help sections contained in the DTCEM help system. These sections are arranged in the same manner as this manual, by menu topic.

Since Help is actually a window, you can use the Windows features to move around. The DTCEM Help system contains hypertext jump topics which move you to other topics and pop-up topics which display definitions or other information. These jump topics and pop-up topics are green and may be selected by clicking on the text.

Keyword

Help is accessible in DTCEM for any number or value in any of the tables. While in a DTCEM table, you may access context specific help for any of the numbers by clicking in the cell and then clicking the right mouse button.

About

Select **About** from the **Help** menu to see information about your version of DTCEM (Figure 71).

Figure 71: DTCEM About Screen

Window Options

The **Window** menu provides options for you to view the data in the document windows on your screen. The features provided in this menu allow you to open, move, size, and arrange many document windows at one time. The basic controls which allow you to size and arrange the windows include restore, minimize, and maximize. These controls are described below.

When you restore a window, you change it to a previous or medium size which you can then move, size, and close. To restore a maximized document window, click the document Restore button in the upper-left hand corner of a maximized document window or choose Restore from the document Control menu. The document Control menu is the menu containing the commands that will open, close, maximize, minimize, or restore a window. You can display the Control menu by clicking on the small rectangular button in the upper left corner of a window or by pressing Alt+space bar. To restore a minimized document, double-click on the document icon or click on a document icon to open the Control menu and choose Restore. A document window is also restored (unless it is minimized) when you tile or cascade windows.

When you minimize a window the window is reduced to an icon allowing you to keep several documents open at the same time. To minimize a restored document, click the minimize arrow (down arrow) in the upper-right hand corner of the document window or choose minimize from the document Control menu.

When you maximize a document window, the document enlarges to fill up the entire document area. To maximize a restored document, click the maximize arrow (up arrow) in the upper-right corner of the document window or double-click on the title bar. To maximize a minimized document, click on a document icon to open the Control menu and choose Maximize.

The Window menu on the menu bar of DTCEM contains the following additional controls which allow you to size and arrange the DTCEM windows:

Cascade

When you have more than one document window open (but not minimized), you can select **Cascade** from the **Window** menu or press Shift+F5 to restore and arrange the open windows. Cascaded windows overlap so that the title bar of each window is displayed. Click on the title bar to view a window's contents.

Tile

When you have more than one document window open (but not minimized), you can select **Tile** from the **Window** menu or press Shift+F4 to restore and arrange the open windows. Tiled windows are arranged on the screen with no overlapping. To work on one of the windows, click on the title bar of the desired window.

Arrange Icons

When you have one or more windows minimized to icons you may wish to arrange the icons so that they are ordered and easy to view. To arrange the icons, select **Arrange Icons** from the **Windows** menu.

Minimize All

To minimize all open windows to icons, select **Minimize All** from the **Window** menu. The minimized icons are displayed on the bottom of the screen.

Restore All

To restore all windows to the maximum size, select **Restore All** from the **Window** menu.

Close All

To close all open windows, select **Close All** from the **Window** menu.

Maximize Current Window

To maximize the current or active window, select **Maximize Current Window** from the **Window** menu, and click on the maximize (up arrow) button in the upper right hand corner. Alternatively, select **Maximize** from the **Control** menu.

Close Current Window

To close the current or active window, select **Close Current Window** from the **Window** menu, and double-click on the upper left hand corner. Alternatively, select **Close** from the **Control** menu.

Toggle Window

To toggle between open windows, click on CONTROL + F6. This is not a menu option in DTCEM however it is a Windows feature and may be used in DTCEM as it may be used in any other Windows program.

Appendix A: Target Efficiencies for Distribution Transformers

Target Efficiencies for 10 kVA Single-Phase Transformers

B Factor	A Factor			
	$0.00 - $1.99	$2.00 - $3.99	$4.00 - $5.99	> $6.00
< $0.50	98.17	98.28	98.23	98.28
$0.50 - $0.99	98.23	98.65	98.45	98.36
$1.00 - $1.49	98.26	98.74	98.90	98.83
$1.50 - $1.99	98.38	98.69	98.71	98.81
> $2.00	NA	98.69	98.93	98.94

Target Efficiencies for 15 kVA Single-Phase Transformers

B Factor	A Factor			
	$0.00 - $1.99	$2.00 - $3.99	$4.00 - $5.99	> $6.00
< $0.50	98.22	98.33	98.28	98.33
$0.50 - $0.99	98.27	98.70	98.50	98.41
$1.00 - $1.49	98.31	98.78	98.95	98.88
$1.50 - $1.99	98.43	98.73	98.75	98.85
> $2.00	NA	98.74	98.98	98.99

Note: Target Efficiencies are expressed at 50 percent load.
Target Efficiencies may change annually.

Target Efficiencies for 25 kVA Single-Phase Transformers

	A Factor			
B Factor	$0.00 - $1.99	$2.00 - $3.99	$4.00 - $5.99	> $6.00
< $0.50	98.49	98.60	98.55	98.60
$0.50 - $0.99	98.54	98.97	98.77	98.68
$1.00 - $1.49	98.58	99.05	99.22	99.15
$1.50 - $1.99	98.70	99.00	99.02	99.12
> $2.00	NA	99.01	99.25	99.26

Target Efficiencies for 37.5-50 kVA Single-Phase Transformers

	A Factor			
B Factor	$0.00 - $1.99	$2.00 - $3.99	$4.00 - $5.99	> $6.00
< $0.50	98.64	98.76	98.71	98.75
$0.50 - $0.99	98.70	99.13	98.92	98.83
$1.00 - $1.49	98.73	99.21	99.38	99.31
$1.50 - $1.99	98.86	99.16	99.18	99.28
> $2.00	NA	99.17	99.41	99.42

Note: Target Efficiencies are expressed at 50 percent load.
Target Efficiencies may change annually.

Target Efficiencies for 75 - 167 kVA Single-Phase Transformers

B Factor	A Factor			
	$0.00 - $1.99	$2.00 - $3.99	$4.00 - $5.99	> $6.00
< $0.50	98.98	99.05	98.90	99.26
$0.50 - $0.99	99.02	99.25	99.17	99.01
$1.00 - $1.49	99.19	99.29	99.32	99.46
$1.50 - $1.99	99.11	99.21	99.34	99.40
> $2.00	NA	99.26	99.36	99.59

Note: Target Efficiencies are expressed at 50 percent load.
Target Efficiencies may change annually.

Appendix B: Glossary

Annualized: The annualized cost of transmission and distribution per kW.

A value: The dollar value, expressed in $/Watt, for the power losses that occur as a result of energizing the transformer. It is multiplied by the core (no-load) losses, in Watts, to form one value to obtain the total owning costs associated with no-load losses.

Avoided Cost: The avoided cost of operating the needed capacity using the base and change case scenarios. This value is equal to the present value difference divided by 1,000 and multiplied by the decrement.

Avoided Cost of Energy (EC): The levelized avoided (incremental) cost for the next kWh produced by the utility's generating units. It consists of a fuel component and a variable operations and maintenance component.

Avoided Cost of Generation Capacity (GC): The incremental cost of adding an additional kW of generating capacity to the utility's system. There may be multiple scenarios for the installation of new capacity, and the avoided cost typically represents the costs (or the savings) from using the least costly scenario.

Avoided Cost of System Capacity (SC): The levelized avoided (incremental) cost of generating transmission and distribution. The avoided cost of system capacity consists of two components: avoided cost of Generation Capacity (GC) and the avoided costs of Transmission and Distribution Capacity (TD). SC = GC + TD

Avoided Cost of Transmission and Distribution Capacity (TD): The incremental cost of adding an additional kW of transmission and distribution capacity to the utility's system. There may be multiple scenarios for the installation of new T&D capacity, and the avoided cost typically represents the costs (or the savings) from using the least costly scenario.

B value: The dollar value, expressed in $/Watt, for the power losses due to electric load on the system served by the transformer. It is multiplied by the winding (load) losses to form one value to obtain the total owning costs associated with the load losses.

Bid Price: The purchase price to acquire the transformer from the manufacturer. Transformer manufacturers provide this information when responding to utility requests for quotation.

Capital Recovery Factor: A factor that, when multiplied by the principal amount of a loan, determines the annual constant payment needed to repay the loan with interest in a given number of years.

Change-out Loading: The percentage of peak loading that the transformer reaches when it is removed from service. Values are usually well over 100%.

Core Losses: Core losses, or no-load losses, are the excitation losses at rated voltage when the transformer is not supplying any load. Core losses occur on all energized transformers and are continuous, independent of load.

Cost of Capital: The composite cost which shows the value of the return to an investor, consisting of a combination of interest rate, bond rate, and/or dividends. It is also known as the minimum acceptable return. The cost of capital is a component of the fixed charge rate.

Decrement: The number of gigaWatt hours which is estimated to predict the change case avoided cost of energy or the number of kW used to predict the change case avoided cost of capacity.

Demand Charge: The demand charge is the per kW of electricity used. These charges typically vary by time of year and peak utility season. In this screen you should enter the average demand charge in dollars per kW per month in the corresponding cream colored box.

Demand Side Management (DSM): The planning, implementation, and monitoring of utility activities designed to encourage consumers to modify the patterns and/or amount of electricity usage, including the timing and level of electricity demand.

Discount rate: The interest rate used to convert future payments into present values.

(EC): The levelized avoided (incremental) cost for the next kWh produced by the utility's generating units. It consists of a fuel component and a variable operations and maintenance component.

Energy Charge: The energy charge is the cost per kWh of electricity used. These charges vary by both time of year, time of day, and amount used (block rates). In this screen you should enter the average energy charge in dollars per kWh.

ENERGY STAR Transformer Program: A voluntary, EPA-sponsored program to recognize electric utilities which make a commitment to purchase high-efficiency distribution transformers. The recognition comes in the form of public outreach campaigns to highlight ENERGY STAR Partners' commitment to saving energy and reducing associated air emissions. In addition, the EPA will help with the development of technical resources to assist utilities' efforts to optimize transformer purchases.

Equivalent Annual Peak Load (PL2): The levelized annual peak load seen by the transformer that is equivalent to an initial peak load with an estimated load growth rate and a maximum allowable load before change out is required.

FCR: The cost of carrying a capital investment, or the annual owning cost of an investment as a percentage of the investment. The components of fixed charge rate include the cost of capital, depreciation on the investment, insurance, and taxes (local, income, etc.).

Fixed Charge Rate (FCR): The cost of carrying a capital investment, or the annual owning cost of an investment as a percentage of the investment. The components of fixed charge rate include the cost of capital, depreciation on the investment, insurance, and taxes (local, income, etc.).

GC: The incremental cost of adding an additional kW of generating capacity to the utility's system. There may be multiple scenarios for the installation of new capacity, and the avoided cost typically represents the costs (or the savings) from using the least costly scenario.

Greenhouse gas: An atmospheric gas which is transparent to incoming solar radiation but absorbs the infrared radiation emitted by the Earth's surface. The principal greenhouse gases are carbon dioxide, methane, and CFCs.

Hours per year (HPY): Hours per year used in the A and B value equations is generally taken as 8,760. However, there may be exceptions, such as transformers connected to irrigation loads. In this case, the transformer may be de-energized a portion of the year, and the value would be less than 8,760.

Inflation Rate: The rate of increase in available currency and credit beyond the proportion of available goods.

Initial Transformer Loading: The peak load percentage that the transformer is serving at beginning of its life.

Intercept: The no load transmission and distribution cost per kW.

Kilowatt (kW): One kilowatt (kW) is equal to 1,000 watts or the absolute meter kilogram per second unit of power equal to the work done at the rate of one absolute joule per second or to the ate of work represented by a current of one ampere under a pressure of one volt and taken as the standard in the United States.

Kilowatt Hour (kWh): A unit of work or energy equal to that expended by one kilowatt in one hour or to 3.6 million joules.

LF: The average energy used for a given period of evaluation (year, month, or day).

LM: A factor used to measure the impact of transmission and distribution system losses on the generation system. For example, if a transformer loses 1,000 kWh/year and the transmission and distribution system losses average 10%, 1,100 kWh (1,000 * (1 + 1.0) will need to be generated to replace these losses.

Load Factor: The average energy used for a given period of evaluation (year, month, or day).

Load losses: See winding losses.

Loss Multiplier (LM): A factor used to measure the impact of transmission and distribution system losses on the generation system. For example, if a transformer loses 1,000 kWh/year and the transmission and distribution system losses average 10%, 1,100 kWh (1,000 * (1 + 1.0) will need to be generated to replace these losses.

LsF: A ratio of the annual average load losses to the peak value of load losses on the distribution transformer.

Minimum acceptable return: The lowest interest rate acceptable to a utility for any investment. It may be equal to a discount rate (the interest the utility would have to pay on a loan) or equal to the interest rate being paid to bondholders, or equal to the utility stock dividend yield (if applicable).

No-load Losses: See core losses.

Peak load growth rate: The rate at which the peak load on the transformer increases each year.

Peak Responsibility Factor (RF): A measure of the diversity of the load on the transformer. It defines the relationship between the transformer peak load and the transformer load at the time of the utility system peak load.

PL2: The levelized annual peak load seen by the transformer that is equivalent to an initial peak load with an estimated load growth rate and a maximum allowable load before change out is required.

Present value (PV) Base Case: The present value cost of operating or purchasing the base case capacity in millions of dollars.

Present value (PV) Change: The present value cost of operating or purchasing the change case capacity in millions of dollars.

Present value (PV) Difference: The difference between the present value costs of operating the base case and change case capacities in millions of dollars or the difference between the present value costs of purchasing the base case and change case capacities in millions of dollars.

Reserve Margin (Safety Factor): The percentage of capacity required to ensure system reliability above and beyond the utility annual peak kW or MW demand. Typical values are between 15 and 20%.

RF: A measure of the diversity of the load on the transformer. It defines the relationship between the transformer peak load and the transformer load at the time of the utility system peak load.

SC: The levelized avoided (incremental) cost of generating transmission and distribution. The avoided cost of system capacity consists of two components: avoided cost of Generation Capacity (GC) and the avoided costs of Transmission and Distribution Capacity (TD). SC = GC + TD

Slope: The marginal, or incremental, cost of installing T&D capacity ($/kW).

Supplier: The name of the manufacturer that provided the particular bid.

TD: The incremental cost of adding an additional kW of transmission and distribution capacity to the utility's system. There may be multiple scenarios for the installation of new T&D capacity, and the avoided cost typically represents the costs (or the savings) from using the least costly scenario.

Total Owning Cost (TOC): The value that shows the overall cost of a transformer purchase. Most utility cost-effectiveness calculations for transformer purchases determine the total owning cost of the transformer by summing the initial purchasing price of the transformer and the equivalent present value of transformer losses using the A and B values.

Transformer: A transformer converts electricity from one voltage to another voltage. Copper or aluminum conductors are wound around a magnetic core to transform current from one voltage to another. Liquid insulation material or air surrounds the transformer core and conductors to cool and electrically insulate the transformer.

Transformer life: The number of years that the transformer is in service. 30 years is generally used for transformer life.

Transformer Loss Factor (LsF): A ratio of the annual average load losses to the peak value of load losses on the distribution transformer.

Transformer Size: Transformers are assigned a nameplate rating, according to a theoretical maximum load which they are designed to handle. One kVA is roughly equivalent to one kW of electricity demand. Thus, a 10 kVA transformer is operating at or above rated load when the demand on the transformer is 10 kW. In practice, transformers can operate at 200 percent of their capacity for limited periods throughout the year. However, many transformers used in residential applications are usually loaded as lightly as 15-20 percent. Distribution transformers range in size from less than 10 kVA to as large as 5,000 kVA.

Transformer Type (mounting): Distribution transformers are mounted either on an overhead pole (outdoor) or on a concrete pad (outdoor or indoor). Since transmission lines traditionally have been run above ground, pole-top transformers comprise the majority of transformers in service on utility distribution systems.

Winding Losses: The winding losses, or load losses, are the I^2R losses of the transformer at 100% nameplate rated load (e.g., the winding losses that occur when a 25 kVA transformer has a 25 kVA load).

Years to reach change out load: The number of years it takes for the transformer, starting at a certain initial transformer loading, and with its peak load growing at a certain rate, for the transformer to reach the change-out loading level. Values will vary due to the sector (e.g., residential, commercial, or industrial) served, economic conditions, and internal utility practices.

Bibliography

Andreas, John C. 1992. *Energy-Efficient Electric Motors*. New York: Marcel Dekker, Inc.

Anshin, V., and E. Minsker. *Assembly of Power Transformers*. Moscow: MIR Publishers.

ANSI/IEEE Standard C57.104-1991. *Guide for Interpretation of Gases Generated in Oil-Immersed Transformers*. Piscataway, N. J.

ANSI/IEEE Standard C57.106-1991. *Guide for Acceptance and Maintenance of Insulating Oil in Equipment*. Piscataway, N.J.

ANSI/IEEE Standard 241-1990. *IEEE Power Distribution Apparatus*. Piscataway, N.J.

ANSI/IEEE Standard C57.120-1991. *Loss Evaluation Guide for Power Transformers*. Piscataway, N.J.

ANSI/IEEE Standard C57.12.00-1993. *IEEE Standard General Requirements for Liquid Immersed Distribution, Power, and Regulating Transformers*. Piscataway, N.J.

ANSI/IEEE Standard C57.91-1995. *IEEE Guide for Loading Mineral-Oil-Immersed Transformers*. Piscataway, N.J.

ANSI/IEEE Standard C110-1986. *IEEE Recommended Practice for Establishing Transformer Capability When Supplying Nonsinusoidal Load Currents*. Piscataway, N.J.

Ashley, Steven. 1983. Low-loss power transformers sought: Researchers use amorphous metal strips in cores. *American Metal Market*, vol. 91, pp. 7–8.

Bailey, Donald J. 1986. Trends in high efficiency distribution transformers for the 1980s. *American Society for Metals Proceedings*, Metals Park, Ohio.

Barnes, P. R., J. W. Van Dyke, B. W. McDonnell, and S. Das. 1996. *Determination Analysis of Energy Conservation Standards for Distribution Transformers*. ORNL-6847. Oak Ridge, Tenn.: Oak Ridge National Laboratory.

Barnes, P. R., J. W. Van Dyke, B. W. McDonnell, S. M. Cohn, and S. L. Purucker. 1995. *The Feasibility of Replacing or Upgrading Utility Distribution Transformers During Routine Maintenance*. ORNL-6804/R1. Oak Ridge, Tenn.: Oak Ridge National Laboratory.

Bean, R. L., N. C. Chackan, Jr., H. R. Moore, and E. C. Wentz. 1959. *Transformers for the Electric Power Industry.* New York: McGraw-Hill.

Beaty, Wayne. 1978. Evaluate transformer losses. *Electric World*, vol. 189, no. 2, pp. 55–56.

Beaty, Wayne. 1994. Transformer designs cut costs and improve operations. *Electric Light and Power*, vol. 72, no. 5, pp. 24–27.

Bonneville Power Administration. 1986. *Transmission and Distribution Efficiency Improvement R&D Survey Project.* DOE/BP-17102-1. Washington, D.C.: United States Department of Energy.

Boyd, Edward. 1980. Modified standard transformers best hedge against losses? *Electric Light and Power*, vol. 58, no. 3, pp. 43–44.

Bozarth, A. M., D. A. Duckett, and W. J. Ros. 1994. *Guide for Evaluation of Distribution Transformers.* Schenectady, N.Y.: GE.

Bush, Rick. 1995. Expect the unexpected. *Transmission and Distribution*, vol. 47, no. 7, p. 6.

Canadian Standards Association. 1994. *Maximum Losses for Distribution, Power, and Dry-Type Transformers.* CSA-C802-94. Toronto, Ontario, Canada.

Christensen, Peter C. 1995. *Retail Wheeling: A Guide for End-Users.* Tulsa, Okla: PennWell Publishing Company.

Curran, P. M. 1990. METGLAS alloy distribution transformers: The conservation opportunity. *American Public Power Association Engineering and Operations Workshop.* Anaheim, Calif.: Allied Signal, Inc.

Curran, P. M. Amorphous alloy distribution transformers: The conservation opportunity. *NARUC Committee on Energy Conservation.* San Francisco, Calif.: Allied Signal, Inc.

Department of Energy. 1993. *Financial Statistics of Major Investor-Owned Electric Utilities.* DOE/EIA-0437 (91). Washington, D.C.: DOE.

Department of Energy, 1993. *Electric Plant Cost and Power Production Expenses.* DOE/EIA-0455 (91). Washington, D.C.: DOE.

Department of Energy. 1993. *Assumptions for the Annual Energy Outlook.* DOE/EIA-0527 (93). Washington, D.C.: DOE.

Department of Energy. 1994. *Annual Energy Outlook, 1994—With Projections to 2010.* DOE/EIA-0383 (94). Washington, D.C.: DOE, p. 58.

Department of Energy. 1996. *Short-Term Energy Outlook.* DOE/EIA-0202 (96/2Q). Washington, D.C.: DOE, p. 20.

Department of Energy. 1996. Procedures for consideration of new or revised energy conservation standards for consumer products. *Federal Register.* 10 CFR part 430, vol. 61, no. 136, pp. 36973–37987.

DeStesse, John, S. B. Merrick, R. C. Tepel, and J. W. Callaway. 1987. *Assessment of Conservation Voltage Reduction Applicable in the BPA Service Region.* DOE/BP-1403-1. Richland, Wash.: Battelle Pacific Northwest Laboratory.

DeStesse, John, 1986. *Customer System Efficiency Improvement Assessment: Description and Examination of System Characterization Data.* PNL-5995. Richland, Wash.: Batelle Pacific Northwest Laboratory.

Dunklin, Philip I. 1996. *Competition in the Utility Industry,* 2d ed. Atlanta, Ga.: Chartwell.

Electric Power Research Institute, Bonneville Power Administration, and Electrotek Concepts. 1995. *Power Quality Workbook for Utility and Industrial Applications.* TR-105500. Palo Alto, Calif.: CPRI.

Electric Power Research Institute. 1994. *Cost-Effectiveness Analysis of Amorphous Core Transformers Using EPRI DSManager.* TR-104246. Palo Alto, Calif.: EPRI.

Electric Power Research Institute. *Low-Loss Amorphous Metal for Transformer Cores.* RP-1290, sheet no. 39, EPRI Information Sheet: Palo Alto, Calif.: EPRI.

Enabnit, Elgin, G., Jr. 1991. Amorphous-core transformers deserve another look. *Transmission and Distribution,* vol. 43, no. 12, p. 8.

Enholm, Gregory B., and J. Robert Malko. 1994. *Electric Utilities Moving into the 21st Century.* Arlington, Va.: Public Utilities Reports, Inc.

Environmental Protection Agency, Apr. 1997. *A Manual for Use with the Distribution Transformer Cost Evaluation Model (DTCEM).* EPA 430-R-96-020. Washington, D.C.: EPA.

Feinberg, R. 1979. *Modern Transformer Practice.* New York: John Wiley & Sons.

Fisher, Arthur. 1983. Unconventional transformer. *Popular Science,* vol. 222, p. 11.

Foder, George M. 1995. ENFORCED efficiency. *Industrial Distribution,* vol. 84, no. 6, pp. 82–84.

Frank, Jerry. 1993. The how and why of *K*-factor transformers. *Electrical Construction and Maintenance,* vol. 92, pp. 78–81.

Flanagan, William M. 1993. *Handbook of Transformer Design and Applications.* New York: McGraw-Hill.

Franklin, A. C., and D. P. Franklin, 1983. *The J&P Transformer Book.* London: Butterworths.

Gebert, L. G., and K. R. Edwards. 1974. *Transformers: Principles and Applications.* Chicago, Ill.: American Technical Publications, Inc.

Girgis, R. S. 1995. Design and performance improvements in power transformers using ribbon cable. *IEEE Transactions on Power Delivery,* vol. 10, no. 2, pp. 869–877.

Gish, Brian. 1996. FERC's Mega-NOPR: The IOUs respond. *Public Utilitiers Fortnightly,* vol. 134, no. 5, p. 37–40.

Gonzalez, D. A., G. Gauger, A. Yerges, and G. Goedde. 1997. Distribution transformer for the 21st century. *CIRED '97 Conference Transactions.* Birmingham, England: The Institution of Electrical Engineers.

Howe, B. 1993. *Distribution Transformers: A Growing Energy Savings Opportunity.* Boulder, Colo.: Tech Update, E Source, Inc.

Jones, James R. 1984. Economic selection of transformers for commercial/industrial facilities. *ICPS 84,* pp. 11–14.

Kennedy, Barry W. 1983. Spotting and cutting distribution losses. *Electric World,* vol. 197, no. 6, pp. 143–144.

Kennedy, Barry W. 1997. Transmission and Distribution Efficiency in a Restructured Industry. *IEEE Power Engineering Review,* vol. 17, no. 6, pp. 6–7.

Kennedy, Barry W., and Patricia Gold. 1991. Distribution system efficiency evaluated via computer. *Transmission and Distribution,* vol. 43, no. 4, pp. 64–67.

Kennedy, Barry W. 1991. Bonneville Power Administration goes amorphous. *Moving to Metglas,* vol. 3, no. 1.

Kennedy, Barry W. 1991. Ultra-efficient transformers under Bonneville Power Administration billing credit program for utility conservation. *IEEE/PES 1991 T&D Conference AMDT Panel Session.* Dallas, Tex.: IEEE.

Kennedy, Barry W., Ed Peterson, and Janelle Schmidt. 1993. Customer rebates for installing amorphous core transformers will save BPA energy. *Innovators with EPRI Technology.* Palo Alto, Calif.: EPRI.

Knight, Denise E. 1990. Amorphous alloy distribution transformers: The conservation opportunity. *NARUC Committee on Electricity*. Los Angeles, Calif.: Allied Signal, Inc.

Lee, Albert C., and Sampat P. Mahesh. 1990. Seven years of operating experience with amorphous metal transformers. *Workshop on amorphous core transformers*. New Delhi, India: GE.

Lowdermilk, L. A., and A. C. Lee. 1987. Five years operating experience with amorphous transformers. *Hard and Soft Magnetic Materials Symposium*. Metals Park, Ohio: ASM International.

Lowdon, E. 1989. *Practical Transformer Design Handboook*. Blue Ridge Summit, Pa.: TAB Professional and Reference Books.

Luborsky, R. E., and J. J. Becker. 1979. Strain-induced anisotropy in amorphous alloys and the effect on torrid diameter magnetic properties. *MAG*, vol. 15, no. 6, pp. 1139–1145.

McDonald, D. 1957. *Power Transformers for High-Voltage Transmission*. Edinburgh: Bruce Peebles & Co., Ltd.

McPherson, G., and R. Laramore. 1990. *Electrical Machines and Transformers*. New York: John Wiley & Sons.

Minsker, E., and V. Anshin. *Assembly of Power Transformers*. Moscow: MIR Publishers.

Morgan, R. B. 1992. Choosing the right transformer, part 2. *Electrical Construction and Maintenance*, vol. 91, no. 5, pp. 67–78.

Morgan, R. B. 1992. Choosing the right transformer, part 3. *Electrical Construction and Maintenance*, vol. 91, no. 11, pp. 22–28.

Morgan, R. B. 1993. Choosing the right transformer, part 4. *Electrical Construction and Maintenance*, vol. 92, pp. 43–46.

Moravek, James M., and Edward Lethert. 1993. The *K* factor: Clearing up its mystery. *Electrical Construction and Maintenance*, vol. 92, pp. 65–71.

Murphy, David E. 1990. Distribution transformer information database from cradle to grave. *Transmission and Distribution*, vol. 42, ISS. 12, pp. 33–36.

Nagel, W. D., 1991. Ultra efficient amorphous substation transformer. *PCI-CON90*, pp. 151–155.

National Electric Manufacturing Association. 1996. *Guide for Determining Energy Efficiency for Distribution Transformers*. No. TP1. Washington, D.C.: NEMA.

Ng, Harry, and Bill Shula. 1987. Transformers with lower losses. *EPRI Journal*, October–November: EPRI, vol. 12, no. 7, pp. 22–27.

Ng, Harry, R. Hasegawa, A. C. Lee, and L. A. Lowdermilk. 1991. Amorphous core distribution transformers. *IEEE Transactions*, vol. 79, no. 11, pp. 1607–1623.

Nickel, D. L., and H. R. Braunstein, 1980. Distribution transformer loss evaluation: Proposed techniques, II—Loud characteristics and system cost parameters. *T-PAS*, pp. 788–811.

Pansini, Anthony J. 1988. *Electrical Transformers and Power Equipment*. Englewood Cliffs, N.J.: Prentice-Hall.

Paula, Greg. 1989. New insulation improves transformer performance. *Electric World*, vol. 203, no. 14, pp. 45–46.

Pavik, D., D. C. Johnson, and R. S. Girgis. 1993. Calculation and reduction of stray and eddy losses in core-form transformers using a highly accurate finite element modeling technique. *IEEE Transactions on Power Delivery*, vol. 8, no. 1, pp. 239–245.

Radtke, Michael. 1991. Failure analysis improves distribution transformer quality. *Transmission and Distribution*, vol. 43, no. 11, pp. 82–88.

Reason, John. 1991. How to buy the best transformer. *Electrical World*, vol. 205, no. 9, pp. 73–84.

Reason, John. 1992. How electric utilities buy quality when they buy transformers. *Electrical World*, vol. 206, no. 5, pp. 49–52.

Reason, John. 1994. Performance improves in small steps. *Electrical World*, vol. 208, no. 6, pp. 48–60.

Reason, John, 1994. Transformer rebuild increases load capacity. *Electrical World*, vol. 208, no. 2, pp. 38–40.

Rice, D. E. Adjustable-speed drive and power rectifier harmonics: Their effects on power system components. *Proceedings of the IEEE PCIC Conference*. IEEE No. PCIC-84-52. New York: IEEE.

Richardson, D. V. 1978. *Rotating Electric Machinery and Transformer Technology*. Reston, Va.: Prentice-Hall.

Rural Electric Administration. *Guide for Economic Evaluation of Distribution Transformers*. REA Bulletin 61-16. Washington, D.C.: United States Department of Agriculture.

Schuler, Joseph F., Jr. 1996. Superconductors help or hype? *Public Utilities Fortnightly*, vol. 134, no. 2, pp. 30–34.

Schuler, Joseph F., Jr. 1996. Power pundits make their pitches. *Public Utilities Fortnightly*, vol. 134, no. 5, pp. 41–42.

Sealey, D. C. 1946. *Transformers Theory and Construction*, Scranton, Pa.: International Textbook Company.

Smith, S. 1985. *Magnetic Components Design and Applications*. New York: Van Nostrand Reinhold.

Tepel, R. C., J. W. Callaway, and J. G. DeSteese. 1987. *Customer System Efficiency Improvement Assessment: Supply Curves for Transmission and Distribution Conservation Options*. PNL-6076. Richland, Wash.: Battelle Pacific Northwest Laboratory.

T&D Special Report. 1991. Redesigned insulation system increases transformer rating, *Electrical World*, vol. 205, no. 3, pp. 10–11.

T&D Special Report. 1993. How to cut losses and tighten margins, get more from T&D equipment. *Electrical World*, vol. 207, no. 10, pp. 56–59.

United Laboratories Standards 1561. *Dry-Type General Purpose and Power Transformers*.

United Laboratories Standards 1562. *Transformers, Distribution, Dry-Type-Over 600 Volts*.

Waggoner, Ray. 1993. Harmonics: A systems related problem. *Electrical Construction and Maintenance*, vol. 92, pp. 22–24.

Walling, R. A., R. K. Hartana, R. M. Reckard, M. P. Sampat, and T. R. Balgie. 1994. Performance of metal-oxide arresters exposed to ferroresonance in padmount transformers. *T-PWRD*, pp. 788–795.

Ward, Daniel. 1990. Evaluating product reliability costs. *IEEE Transactions, 89Sm779-OPWRD.* vol. 180, pp. 724–729.

Western Area Power Administration. 1986. Distribution System Loss Evaluation Manual, Denver, Colo.: WAPA.

Whitley, Dorman W. 1986. Type testing of amorphous metal distribution transformers. 86 T&D 555-7. *IEEE/PES 1986 Transmission and Distribution Conference.* Anaheim, Calif.: IEEE.

Wolf, Robert F. 1982. Save dollars on transformer losses. *Electric World,* vol. 196, no. 2, pp. 119–120.

Zackerison, Harry B. 1983. *Energy Conservation Techniques for Engineers.* New York: Van Nostrand Reinhold.

Index

About the Author

Barry W. Kennedy, an electrical engineer with the
Bonneville Power Administration in Oregon, specializes in
improving electric transmission and distribution efficiency.